MOLECULAR BIOLOGY
INTELLIGENCE
UNIT

Eukaryotic DNA Damage Surveillance and Repair

Keith W. Caldecott, Ph.D.
Genome Damage and Stability Centre
University of Sussex
Falmer Brighton, U.K.

LANDES BIOSCIENCE / EUREKAH.COM
GEORGETOWN, TEXAS
U.S.A.

KLUWER ACADEMIC / PLENUM PUBLISHERS
NEW YORK, NEW YORK
U.S.A.

EUKARYOTIC DNA DAMAGE SURVEILLANCE AND REPAIR

Molecular Biology Intelligence Unit

Landes Bioscience / Eurekah.com
Kluwer Academic / Plenum Publishers

Printed in the U.S.A.

Kluwer Academic / Plenum Publishers, 233 Spring Street, New York, New York, U.S.A. 10013
http://www.wkap.nl/

Please address all inquiries to the Publishers:
Landes Bioscience / Eurekah.com, 810 South Church Street
Georgetown, Texas, U.S.A. 78626
Phone: 512/ 863 7762; FAX: 512/ 863 0081
www.Eurekah.com
www.landesbioscience.com
Many of the figures in this book can be viewed in color at www.Eurekah.com.

Eukaryotic DNA Damage Surveillance and Repair, edited by Keith Caldecott, Landes / Kluwer dual imprint / Landes series: Molecular Biology Intelligence Unit

ISBN: 0-306-47987-7

Library of Congress Cataloging-in-Publication Data

Caldecott, Keith W.
Eukaryotic DNA damage surveillance and repair / Keith W. Caldecott.
p. ; cm.
Includes index.
ISBN 0-306-47987-7
1. DNA damage. 2. DNA repair. 3. Eukaryotic cells. I. Title.
[DNLM: 1. DNA Damage--physiology. 2. DNA Repair--physiology. 3. Eukaryotic Cells--metabolism. QH 465 C146e 2003]
QH465.A1C34 2003
572.8'6--dc22

2003022447

CONTENTS

Preface vii

1. **DNA Damage Surveillance in *Saccharomyces cerevisiae*** 1
Geraldine W-L. Toh, Stefano Maffini and Noel F. Lowndes
Introduction 1
Sensing DNA Damage: A Complex Problem 4
From Checkpoint Genes to Protein Complexes 6
The Rad24 Epistasis Group: RFC-Like and PCNA-Like Complexes 7
Mec1/Ddc2: Signal Transduction and More 10
The Mre11 Complex: DSBs and Checkpoint Signalling 12
The S Phase Checkpoint 15
Checkpoints and Genome Surveillance 16
Focal Recruitment of Checkpoint and Repair Complexes to Damage Sites 17
The Chromatin Connection 18
Concluding Remarks 20

2. **Mammalian Base Excision Repair** 27
Grigory L. Dianov, Sarah L. Allinson, Helen Budworth and Kate Sleeth
Abstract 27
Introduction 27
Base Excision Repair Proteins 28
Coordination of Base Excision Repair 36

3. **Detection and Repair of DNA Single-Strand Breaks** 49
Keith W. Caldecott
Abstract 49
The Origin and Structure of DNA Single-Strand Breaks 49
The Threat of Single-Strand Breaks to Genetic Stability and Cell Death 50
Polypeptides Involved in DNA Single-Strand Break Repair 50
PARP-1, XRCC1 and the Detection of SSBs 50
AP Endonuclease-1 (APE1) 53
Polynucleotide Kinase (PNK) 54
DNA Polymerase-β (Polβ) 54
DNA Ligase III 55
SSBR Protein Complexes; Pre-Formed or Assembled at Breaks? 56
SSBR Pathways and the Cell Cycle 57

4. **The Detection and Repair of Nucleotide Damage and Coupling to Major Cellular Regulatory Processes in Mammalian Cells** 63
James E. Cleaver
Abstract 63
Introduction 63
Types of Damage Repaired by NER 66
Recognition of Damage in Nontranscribing Regions of the Genome 67

Recognition of Damage in Transcriptionally Active Regions of the Genome 68
The Core Excision Process Common to GGR and TCR 70
Ubiquitination and Proteolysis in DNA Repair 73
Inducible Responses to UV Irradiation 74
Kinase Activation and Cellular Response to UV Damage 75
Replication of Damaged DNA 76
Differentiation and Tissue-Specific Expression of DNA Repair Genes 78
Polymorphisms in DNA Repair Genes 78

5. **Mammalian DNA Mismatch Repair** **90**
Helen M.R. Robinson and Robert Brown
Abstract 90
What Is DNA Mismatch Repair 90
The Mismatch Repair Mechanism 91
The Mismatch Repair Proteins 93
The MutS Homolog Proteins 96
Proteins That Interact with the MuL and MutS Homologues 98
Apoptosis and Drug Resistance 101
MMR and Apoptotic Pathways 102
Conclusions 103

6. **Origin, Recognition, Signaling and Repair of DNA Double-Strand Breaks in Mammalian Cells** **107**
Larry H. Thompson and Charles L. Limoli
Abstract 107
Replication-Independent Double-Strand Breaks (DSBs) 108
Replication Associated DSBs (One-Sided Breaks) 121

7. **The Mechanism of DNA Non-Homologous End-Joining: Lessons Learned from Biophysical, Biochemical and Cellular Studies** **146**
P.A. Jeggo
Abstract 146
Introduction 146
Nature of the DNA Ends to Be Rejoined 147
Structure and Biochemical Analysis of the Core NHEJ Proteins 148
Additional Proteins Implicated in DNA NHEJ 150
Further Proteins Interacting with Core NHEJ Components 151
Model for NHEJ and Potential Additional Interacting Proteins 152

Index **159**

EDITOR

Keith W. Caldecott, Ph.D.
Genome Damage and Stability Centre
University of Sussex
Falmer Brighton, U.K.
k.w.caldecott@sussex.ac.uk
Chapter 3

CONTRIBUTORS

Sarah L. Allinson
Radiation and Genome Stability Unit
Medical Research Council
Harwell, Oxfordshire, U.K.
Chapter 2

Robert Brown
Cancer Research U.K. Department
Medical Oncology
Beatson Laboratories
Glasgow University
Glasgow, U.K.
Chapter 5

Helen Budworth
Radiation and Genome Stability Unit
Medical Research Council
Harwell, Oxfordshire, U.K.
Chapter 2

James E. Cleaver
Departments of Dermatology
and Pharmaceutical Chemistry
Auerback Melanoma Laboratory
UCSF Cancer Center
University of California
San Francisco, California, U.S.A.
Chapter 4

Grigory L. Dianov
Radiation and Genome Stability Unit
Medical Research Council
Harwell, Oxfordshire, U.K.
Chapter 2

P.A. Jeggo
Genome Damage and Stability Centre
University of Sussex
Brighton, U.K.
Chapter 7

Charles L. Limoli
Department of Radiation Oncololgy
University of California San Francisco
San Francisco, California, U.S.A.
Chapter 6

Noel F. Lowndes
Genome Stability Laboratory
Department of Biochemistry
National Centre for Biomedical
Engineering Science
National University of Ireland
Galway, Ireland
Chapter 1

Stefano Maffini
Genome Stability Laboratory
Department of Biochemistry
National Centre for Biomedical
Engineering Science
National University of Ireland
Galway, Ireland
Chapter 1

Helen M.R. Robinson
Cancer Research U.K. Department
Medical Oncology
Beatson Laboratories
Glasgow University
Glasgow, U.K.
Chapter 5

Kate Sleeth
Radiation and Genome Stability Unit
Medical Research Council
Harwell, Oxfordshire, U.K.
Chapter 2

Larry H. Thompson
Biology and Biotechnology Research Program
Lawrence Livermore National Laboratory
Livermore, California, U.S.A.
Chapter 6

Geraldine W-L. Toh
Genome Stability Laboratory
Department of Biochemistry
National Centre for Biomedical Engineering Science
National University of Ireland
Galway, Ireland
Chapter 1

PREFACE

The field of eukaryotic DNA repair has experienced an immense increase in exposure and popularity in the last ten years, not least because of the discovery of direct links between this process and human genetic diseases. Without doubt the most intensively studied area has been the link between DNA repair and other cellular responses to DNA damage, and in particular the control of cell cycle progression. The interplay between DNA repair enzymes and cell cycle checkpoint proteins is a rapidly emerging area in which we have learned much in recent years, but of which we most certainly have much more to learn. Perhaps the area of most topical interest, and also of which we know least, is the molecular mechanism by which DNA damage is first detected and then signaled to the DNA repair, cell cycle checkpoint, and when appropriate, apoptotic machinery. In this book experts in their fields review each of the major types of DNA repair process, with particular emphasis placed on the initial steps of DNA damage surveillance and signaling.

Keith W. Caldecott, Ph.D.

Chapter 1

DNA Damage Surveillance in *Saccharomyces cerevisiae*

Geraldine W-L. Toh, Stefano Maffini and Noel F. Lowndes

Introduction

Damage to genomic DNA is a significant and constant problem for eukaryotic cells. Environmental mutagens, ionizing radiation and even oxidative metabolism within the cell, can all result in damage to DNA, a problem compounded by DNA's intrinsic biochemical instability.[1] DNA damage triggers a wide range of cellular responses, including altered gene expression, cell cycle arrest and stimulation of DNA repair (reviewed in ref. 2). Left unrepaired, base damage or modification can lead to mutations, while other types of lesions such as strand-breaks and cross-links interfere with processes such as DNA replication, gene transcription and chromosome segregation. The ability to deal with spontaneous or environmentally induced DNA damage is as critical to both genome stability and cell viability as accurate DNA replication or proper chromosomal segregation. Consequently, eukaryotic cells have evolved surveillance mechanisms, known as DNA damage checkpoints that monitor genome integrity and coordinate various aspects of the damage response, including cell cycle progression, gene transcription and DNA repair. In this review, we focus primarily on DNA damage sensing in the budding yeast, *Saccharomyces cerevisiae*, with reference to fission yeast, *Schizosaccharomyces pombe*, and mammalian cells, to underscore common and diverse features in eukaryotic genome surveillance. See Table 1.

A DNA damage checkpoint is a signal transduction pathway that senses the presence of DNA damage and transmits a signal to downstream effectors to execute the various cellular responses to DNA damage. Checkpoints signal the presence of either abnormal or incompletely assembled cellular structures that may be caused by the cell's own error-prone internal machinery (e.g., replication fork collapse), the metabolic environment (e.g., hydrolysis and oxidation of DNA) or may be prompted by external agents (e.g., carcinogens). Damage checkpoints attenuate the mutagenicity, genomic instability and cell lethality that can result from DNA damage, by delaying cell cycle progression, increasing the transcription of genes involved in DNA repair, as well as directly modulating DNA repair enzyme activity. In mammalian cells, several damage checkpoint pathway components also function as tumour suppressors and defects in checkpoint signalling pathways are frequently associated with cancer.[2] Checkpoint controls appear to be highly conserved. The cellular responses to damage (e.g., cell cycle arrest, inhibition of DNA replication) are found in all eukaryotic cells that have been examined and genes that mediate these responses are, with few exceptions, also highly conserved.

Eukaryotic DNA Damage Surveillance and Repair, edited by Keith W. Caldecott.

Table 1. Sensors and transducers involved in DNA damage surveillance in **S. cerevisiae**

Functional Class	Protein or Protein Complex	Ascribed Functions	References	Related Genes in Other Species
Sensors of DNA damage	Rad24 Rfc2-5	RFC-like complex; loading of PCNA-like complex onto DNA. Damage sensing and/or recruitment of other checkpoint proteins. Required for G1, S & G2 checkpoints.	(22, 46, 47, 72, 130, 131)	Rad17 (*S. pombe*), Rad17 (human)
	Rad17 Ddc1 Mec3	PCNA-like heterotrimeric complex. Damage sensing and/or recruitment of other checkpoint proteins. Required for G1, S & G2 checkpoints.	(49, 132)	Rad9 (*S. pombe*), Rad9 (human) Rad1 (*S. pombe*), Rad1 (human) 3rd member of complex in humans & *S. pombe* is Hus1 which does not appear to be related to Mec3.
	Rad9	BRCT-containing protein. Damage sensor/mediator of damage checkpoint. Activation of Rad53.	(23, 26, 34, 35, 133, 134)	Crb2/Rhp9 (*S. pombe*) BRCA1, 53BP1, Mdc1 (mammalian)
	Mre11 Rad50 Xrs2	Sensing and signalling of double-strand breaks in all phases of cell cycle. Also implicated in DSB repair.	(94, 96, 99-101)	Mre11 (human), Rad32 (*S. pombe*) Rad50 (human), Rad50 (*S. pombe*) Nbs1 (human)
Replication stress checkpoint	Pol2	DNA polymerase epsilon. Intra-S checkpoint.	(33)	None?
	Mrc1	Transduces replication stress signals through Rad53 (intra-S checkpoint). Functions analogously to Rad9 as an adaptor protein, interacting with Tel1 and Rad53 to signal replication stress.	(113)	Mrc1 (*S. pombe*)

continued on next page

Table 1. Continued

Functional Class	Protein or Protein Complex	Ascribed Functions	References	Related Genes in Other Species
	Dpb11	BRCT protein. Interacts with Pol2 and Drc1. Dpb11 and Drc1 interact and control the Rad53-dependent response to damage and replication blocks in S phase.	(135-137)	Rad4/Cut5 (*S. pombe*), TopBP1 (human)
	Drc1			Drc1 (*S. pombe*)
	Sgs1	RecQ-like DNA helicase associated with Pol ε. Colocalises with and acts upstream of Rad53 in S phase checkpoint.	(110)	BLM, WRN (human), Rhq1/ Rad12 (*S. pombe*)
	Srs2	DNA helicase. Required for proper Rad53 activation in intra-S checkpoint. Also required for recovery from damage checkpoint-mediated arrest.	(9, 138)	Srs2
Signal transducers	Mec1	PI3K-like kinase (PIKK). Sensing, transduction and amplification of the damage signal.	(5, 6, 60, 63, 74, 79, 80, 112)	Rad3 (*S. pombe*), ATR (human)
	Ddc2/Lcd1	Coiled-coil protein. Binds to and may regulate activity of associated PIKK.		Rad26 (*S. pombe*), ATRIP (human)
	Tel1	PIKK. Role in telomere maintenance. Partial overlap in function with Mec1 in damage checkpoint. In S phase, functions with Mre11 complex in signalling unprocessed DSBs.	(74, 95)	Tel1 (*S. pombe*), ATM (human)
	Rad53	Required for both DNA damage and replication block checkpoints in budding yeast.	(32, 34, 35, 71, 77, 105, 139)	Chk2 (human), Cds1 (*S. pombe*)
	Chk1	Both Rad53 and Chk1 phosphorylate downstream targets that control replication and cell cycle progression.	(32)	Chk1 (human and *S. pombe*).

Checkpoints were originally defined as control mechanisms that impose order or enforce dependency of events within the cell cycle. In the past, checkpoints were assumed to act as barriers to progression through the cell cycle, permitting only those cells that successfully complete an earlier stage of the cell cycle to pass through to the next one; for example, in eukaryotic cells completion of DNA synthesis is a necessary condition for entry into mitosis. However, it is more correct to think of the DNA damage checkpoint as signal transduction pathways responsible for damage-dependent control, not only of cell cycle progression, but also other, perhaps more critical, aspects of the DNA damage response—notably repair and, in higher eukaryotes, apoptosis. Like other signal transduction pathways, DNA damage checkpoint pathways involve sensors, transducers and effectors. Sensors monitor the DNA for the presence of abnormalities, by interacting directly or indirectly with DNA lesions and then initiate transduction of the checkpoint signal. The transducers further transmit and amplify this signal, while the effectors control the biological consequences of triggering the pathway.

Many of the signal transducers are protein kinases that act by phosphorylating other kinases and/or downstream target proteins, thus triggering a biochemical cascade in response to DNA damage. The practical distinction between 'sensors' and 'transducers' is likely to be blurred, since such 'transducer' kinases as Mec1^{sc}, Rad3^{sp} and ATM are capable of sensing certain types of DNA damage independently of so-called 'sensor' proteins.[3-7]

DNA damage checkpoint pathways also share with other signal transduction pathways the phenomenon of down-regulation of the signal, or adaptation, in the continuous presence of the initiating signal.[8] Moreover, once damage has been successfully repaired, a checkpoint pathway-dependent recovery process is initiated, allowing previously proliferating cells to resume growth.[9,10]

Sensing DNA Damage: A Complex Problem

In order for the checkpoint response to be initiated, DNA damage must somehow be sensed. Monitoring the genome for DNA damage is a hugely complex and demanding task, because of the myriad types of DNA lesion that occur in cells. For example, 20 types of oxidative lesions alone have been described; in addition, there are the numerous types of photoproduct caused by UV light, as well as double-strand breaks (DSBs) caused by ionizing radiation or radiomimetic chemicals.[1] This would seem to demand enormous sensitivity and specificity on the part of the damage sensing apparatus.

How, then, are all these physically dissimilar lesions sensed? Do cells possess a battery of sensors, each specific for an individual lesion, or are there a number of 'adaptor' proteins that can interact with different types of damage? Many DNA repair pathways possess components that bind preferentially to specific types of lesions. By interacting with checkpoint proteins, such components might enable the presence of these lesions to be signalled to the checkpoint machinery. For example, the human XPA (Xeroderma pigmentosum group A) protein, a component of the nucleotide excision repair (NER) pathway, that can selectively recognise UV-induced photoproducts,[11-13] also appears to be required for aspects of the G1 checkpoint in mammalian cells. Similar results have been obtained with the yeast equivalent NER component, Rad14.[14] Another example of a known DNA repair protein that also functions in checkpoint regulation is the Mre11 complex, implicated in DNA double-strand break (DSB) repair. It has been shown to interact with DSBs and is known to be involved in checkpoint signalling of this type of lesion.[15]

DNA repair complexes may not only have direct roles in signalling the presence of lesions to the checkpoint machinery, but may also facilitate signalling by modifying different types of primary lesions to a common structure or structures, e.g., through processing by specific repair pathways, or as a result of DNA replication. Recognition of such structures might then trigger a checkpoint response. See Figure 1. For example, it has been proposed that single-stranded

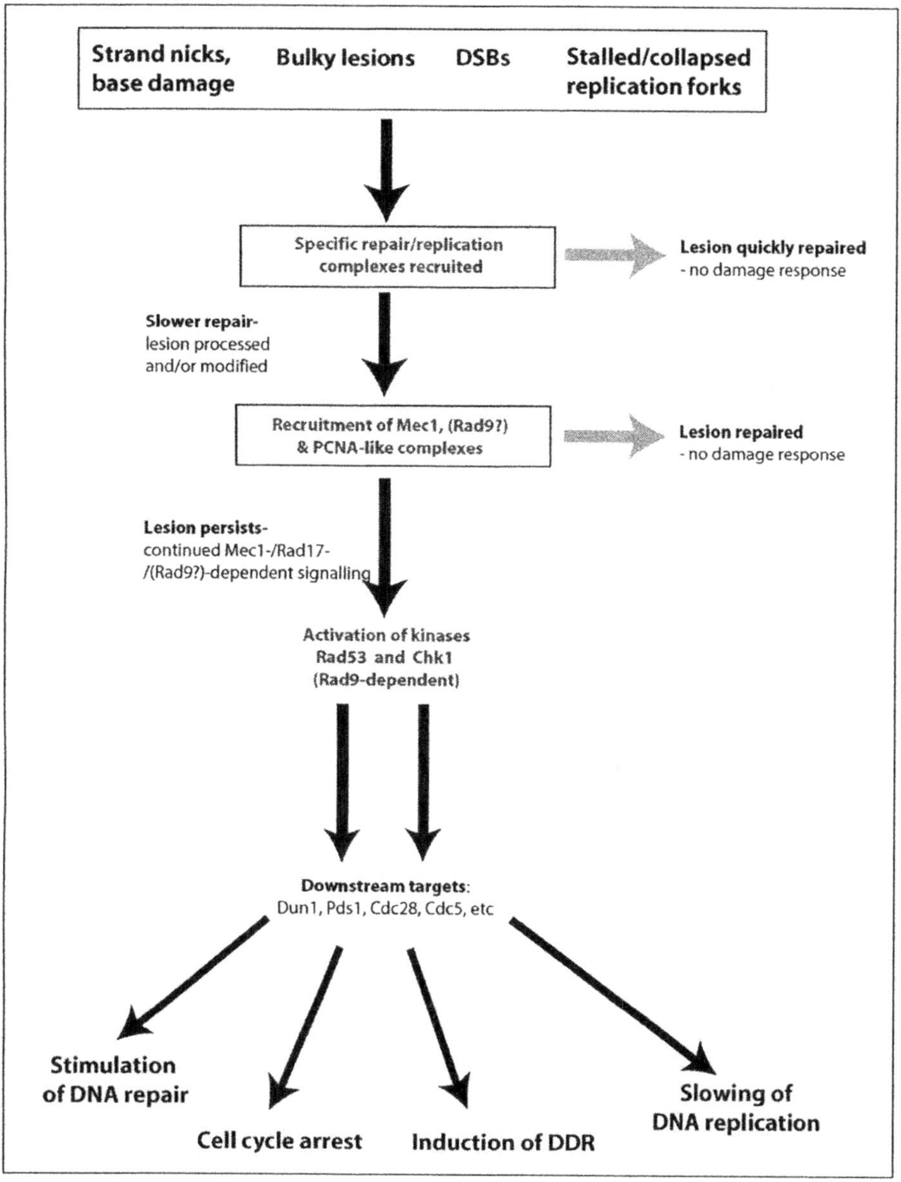

Figure 1. Sensing, signalling and repair of DNA damage. Initial detection of DNA damage by specific repair or replication-associated protein complexes recruited to the lesion. Most lesions are repaired quickly (see text and ref. 68 while slower repair results in lesion processing/modification. Initial recruitment of Mec1 to damage sites allows it to phosphorylate 'local' targets, e.g., Ddc2, H2A and perhaps Rad9; the RFC-like and PCNA-like complexes may also be independently recruited. If repair is successful, activation of the Mec1-dependent damage signalling cascade may be avoided. Checkpoint and repair factors (including the RFC-like and PCNA-like complexes) accumulate at difficult-to-repair lesions, contributing to a sustained and elevated level of Mec1-dependent checkpoint signalling. This results in activation of the transducer kinases Rad53 and Chk1, and phosphorylation of damage checkpoint effectors, and leads to downstream consequences, including cell cycle arrest, transcription, and increased repair.

DNA (ssDNA), generated at lesions, might be the common structure that triggers the Mec1/Rad53-dependent checkpoint response.[9,16-18]

In S phase cells, the process of ongoing DNA synthesis may create further complications for the damage sensing apparatus. In addition to DNA damage, the S phase checkpoint pathway also responds to conditions that interfere with DNA replication, e.g., depletion of nucleotide precursors. Several proteins or protein complexes have been implicated in the initiation of the S phase checkpoint signal, for example DNA Pol ε, Dpb11, RPA, Chl12/Rfc2-5 and Rad24/Rfc2-5, Tof1, Mrc1 and Mec1/Ddc2.[19,20] The precise roles of these proteins in the S phase checkpoint remain unclear, no doubt because most of them also appear to be essential for normal replication. Specific loss-of-function mutants will therefore be required to determine their roles in checkpoint regulation.

In addition to detecting the presence of lesions, the damage sensing apparatus is also likely to be responsible for 'checking' on the progress of lesion repair or processing. Such dynamic monitoring would allow damage signalling to be coordinated at each stage with repair and other critical cellular processes and might provide a way of ensuring that the response is appropriate to the severity of the damage.

From Checkpoint Genes to Protein Complexes

What are the biochemical functions of checkpoint proteins and how are they regulated? Genetic studies originally identified a number of genes required for the DNA damage checkpoint in budding yeast.[17,21] Further work established a genetic order of function and defined pathways involved in checkpoint-mediated responses. For example, two 'branches' of the DNA damage checkpoint, defined by *RAD9* and the *RAD24* epistasis group, were found to function additively in sensing damage outside of S phase.[22]

Rad9: Signalling and Sensing?

Fifteen years ago, a pioneering budding yeast study identified the prototypical DNA damage checkpoint gene, *RAD9*.[23] *RAD9* contributes to transient cell cycle delays throughout the cell cycle[22-25] and loss of this gene impairs checkpoint-induced cell cycle arrest and increases genomic instability.[23,26-28] Budding yeast Rad9 is a large, low abundance, nuclear phosphoprotein that shows little homology to other proteins, except in its C-terminal domain. This domain contains two BRCT (*BRCA1* carboxyl terminus) motifs that are essential for Rad9 function and mediate oligomerisation of Rad9 both before and after DNA damage.[29] The Rad9 protein becomes hyperphosphorylated following DNA damage and this phosphorylation is dependent on the PIKKs (PI3K-like kinases) Mec1 and Tel1.[28,30,31]

Rad9 in turn mediates the Mec1/Tel1-dependent activation of the parallel checkpoint effector pathways involving the checkpoint protein kinases Rad53 and Chk1.[32,33] The damage-induced Mec1/Tel1-dependent hyperphosphorylation of Rad9 allows its association with, and activation of, Rad53.[28,30,31,34,35] Rad53 contains two phospho-residue binding FHA (Forkhead-associated) domains, both of which are thought to contribute to the interaction with, and activation by, the hyperphosphorylated form of Rad9.[30,34,36] The major damage-induced Rad9 phosphorylation sites are required both for the direct interaction of phosphorylated Rad9 with Rad53 and for subsequent Rad53 activation. These phosphorylation sites are not, however, required for DNA damage-induced phosphorylation of Chk1, suggesting that Rad9 regulates these two effector pathways through distinct mechanisms.[34]

Recent genetic and biochemical analyses have revealed how Rad9 may function in activating Rad53. Rad9 exists in two distinct soluble complexes—a large ≥850 kDa complex containing hypophosphorylated Rad9 and a smaller 560 kDa complex, containing hyperphosphorylated Rad9, as well as Rad53. Following DNA damage, remodelling of the large Rad9 complex results in formation of the smaller complex and this change correlates with Rad9 hyperphosphorylation.[35] While bound to the smaller Rad9 complex, Rad53 molecules may be brought into close proximity, facilitating trans-autophosphorylation and full activation of the protein. Activated Rad53 is then released from the complex. Consistent with this model, site-directed mutagenesis has identified multiple sites within Rad9 that are targets for PIKK phosphorylation in response to damage and that are required for Rad9-dependent activation of Rad53.[34] Individual phosphorylation sites were found to support partial levels of Rad9 phosphorylation and Rad53 interaction, but no single phosphorylation site was capable of fully sustaining either function. This suggests that multiple PIKK target sites are together responsible for mediating the Rad9-Rad53 interaction. These observations are consistent with a model in which adsorption of multiple Rad53 molecules occurs, via their FHA domains, during Rad9-mediated catalysis of Rad53 activation.[35]

Both forms of Rad9 complexes also contain the Hsp70 family chaperone proteins Ssa1 and/or Ssa2, suggesting a role for these proteins in checkpoint regulation.[37] Consistent with this possibility, genetic experiments indicate a redundant role for the *SSA1* and *SSA2* genes in survival, G2/M checkpoint regulation and phosphorylation of both Rad9 and Rad53 after UV irradiation. It remains to be determined whether the Ssa1/2 proteins have a function in the remodelling of the hypophosphorylated Rad9 complex after DNA damage.

In addition to its role in transducing the damage signal, Rad9 has been predicted on the basis of genetic analyses to function in the sensing or processing of damage,[17,38] although such a role has not yet been directly demonstrated. Furthermore, neither the large nor the small Rad9 complex shows appreciable binding to naked DNA (C. Gilbert & N. Lowndes, unpublished data). However, this does not preclude the possibilities that Rad9 could be interacting with DNA lesions indirectly, via 'adaptor' proteins or modified chromatin structures. This would remove the requirement for direct contact with DNA. Additionally, it should be noted that both the large and small Rad9 complexes are found in soluble nuclear fractions and nothing is known about the biochemistry of Rad9 at or in the vicinity of the DNA lesions themselves. New technologies will be required to fully investigate the potential role of Rad9 in sensing DNA lesions. An interesting possibility is that Rad9 could be sensing damage indirectly via perturbations to chromatin structure. Consistent with this, a proportion of the total Rad9 can be copurified with chromatin (N.F. Lowndes, unpublished data). A number of recent studies on the effects of disrupting histone acetylation suggest a possible connection between chromatin and the damage checkpoint.[3,39-41] Furthermore, a human equivalent of Mec1, ATM, has recently been shown to be partially activated by treatments that disrupt normal chromatin structure, in the absence of any actual DNA damage.[3]

Additional support for a role for Rad9 in damage sensing comes from recent studies on GFP-tagged budding yeast checkpoint proteins. Rad9, along with a number of other checkpoint proteins, localises to damage-induced nuclear structures, termed foci, that are formed following HO endonuclease-induced DNA double-strand breaks,[5,6] *cdc13-1* induced telomeric damage[5,6] and phleomycin treatment (I. Dobbie, pers. commun.). It will be interesting to investigate, using an approach such as chromatin immunoprecipitation, the kinetics with which Rad9 translocates to DNA lesions. Purification and characterisation of chromatin-bound forms of Rad9 may also help resolve the issue of Rad9 involvement in the damage sensing process.

The Rad24 Epistasis Group: RFC-Like and PCNA-Like Complexes

The *RAD24* epistasis group includes *RAD17*, *MEC3* and *DDC1*, as well as *RAD24* itself. In yeast, these proteins are required for the DNA damage checkpoint in all phases of the cell cycle. It was proposed, based on genetic and biochemical analyses, that these proteins act at an early step in the DNA damage signalling pathway, perhaps as initial sensors of DNA damage. Recent structural and biochemical studies in yeast and human cells have provided some insight into the role of these complexes, but have also raised many more questions.

Rad24 and its homologues in other species have been shown to have homology with all five subunits of RFC (replication factor C),[42] while Rad17, Mec3, Ddc1 and their homologues all exhibit limited homology throughout their lengths with PCNA (proliferating cell nuclear antigen), another replication factor.[43,44] During DNA replication, RFC recognises the primer-template junction and recruits PCNA, onto the DNA, where the latter functions as a "sliding clamp", allowing processive DNA synthesis by polymerases δ and ε.[45]

Rad24 was found to interact with one of the small subunits of the RF-C complex, Rfc5.[46] It was subsequently shown that Rad24 binds to all four small RF-C subunits (Rfc2-5), but not to the large subunit, Rfc1.[47] Green et al thus proposed that Rad24 might form an 'alternative' clamp-loading complex with RFC proteins, in which the large subunit, Rfc1, is replaced by Rad24. The predicted structural similarity between the individual proteins of the Rad17-Mec3-Ddc1 complex and PCNA led to the proposal that the former complex may form a PCNA-like ring structure and function as a 'sliding clamp' during recognition and/or processing of DNA damage.[48,49] It was further hypothesised that the Rad24/RFC putative clamp-loading complex might load the PCNA-like Rad17-Mec3-Ddc1 complex onto damaged DNA, in much the same way that the Rfc1/Rfc2-5 complex loads PCNA homotrimers during DNA replication.

Long-awaited confirmation of this structural model came from transmission electron microscopy (TEM) studies of the human clamp and clamp-loader complexes.[50-52] Direct comparison of TEM images of the two sets of clamp and clamp-loader complexes (PCNA, RFC and Rad9-Hus1-Rad1, Rad17-RFC) indicated that Rad9-Hus1-Rad1 and Rad17-RFC are structural orthologues of PCNA and RFC respectively. (Note the unfortunate nomenclature; human Rad9 and Rad17 are distinct from yeast Rad9 and Rad17. See Table 1). Analyses of the molecular architecture of the two complexes showed that the Rad9-Hus1-Rad1 complex has a characteristic ring structure indistinguishable from that of PCNA in shape and size. In addition, the hRad17-RFC2-5 complex was found to be oval in structure and 26 nm by 22 nm in size with a cleft, reminiscent of the structure of RFC. Furthermore, the purified and reconstituted human Rad9-Hus1-Rad1 (PCNA-like) complex was found to be composed of equimolar amounts of Rad9, Hus1 and Rad1 proteins and to exhibit a native molecular mass of 100 kDa, consistent with a trimeric complex. When human Rad17 was coexpressed with the four small subunits of RFC in insect cells, these proteins formed a complex of 240 kDa that displayed DNA binding, ATPase activity and binding to its predicted target, the Rad9-Hus1-Rad1 complex.

Given the structural similarities of the yeast and human checkpoint complexes to PCNA and RFC, it has been hypothesized that in the DNA damage response a similar loading of the 'clamp' by the 'clamp-loader' complex might occur[48] and perhaps be followed by some sort of processive movement along the DNA duplex. Direct biochemical evidence now exists that the yeast and human PCNA-like and RFC-like complexes can function as clamp and clamp-loader, in a manner analogous to their replicative counterparts.[53,54]

Bermudez et al have shown that the human Rad17-RFC complex binds to the Rad9-Hus1-Rad1 complex in an ATP-dependent manner in vitro, an interaction mediated mainly by the binding of Rad9 to Rad17. Additionally, Rad17-RFC binding to DNA was independent of both ATP and the Rad9-Rad1-Hus1 complex. In contrast, recruitment of Rad9-Hus1-Rad1 complex to DNA required Rad17-RFC and was dependent on the presence of ATP, although not on ATP hydrolysis. Electron microscopy studies also provided visual evidence that loading of Rad9-Hus1-Rad1 by Rad17-RFC onto nicked plasmid DNAs can occur.[54] More recently, Majka and Burgers have used surface plasmon resonance (SPR) to measure in vitro interactions of the budding yeast Rad17/Mec3/Ddc1 and Rad24-RFC complexes with each other and with a range of DNA substrates.[53] Based on the SPR analysis and on in vitro interaction studies, they concluded that ATP binding by Rad24-RFC allows it to load the Rad17/Mec3/Ddc1 complex onto DNA and that its hydrolysis is required to release Rad17/Mec3/Ddc1 from Rad24-RFC. Although partially double-stranded DNA with a 3' primer/template junction was the preferred substrate for Rad17/Mec3/Ddc1 loading, interactions with 5' primer/template junctions and dsDNA ends were also detected. Further extending the parallels with the replicative clamp/clamp-loader system, once loaded, the Rad17/Mec3/Ddc1 complex appears to be able to slide along naked plasmid DNA.[53] Majka and Burgers suggest that this ability may allow the complex to 'scan' the DNA for damage, either alone or in conjunction with other damage recognition/processing factors. Whether this can occur in the context of chromatin and whether it requires chromatin remodelling activities, as seems likely, remains to be determined.

Despite the confirmation of these aspects of the clamp/clamp-loader model, several questions about these complexes remain. At least three versions of RFC-like complexes are known in eukaryotic cells, each with specialised functions: RFC for replication, Rad17-RFC for the damage response and Ctf18/Chl12-RFC for the establishment of sister chromatid cohesion.[55-57] There are also some indications that alternative damage-associated PCNA-like complexes may exist, consistent with genetic studies which attribute overlapping but not identical functions to the three members of the Rad17/Mec3/Ddc1 complex.[58,59] It is not known whether the different forms of these complexes are always present as distinct entities, or if a more dynamic interplay exists, for instance, competition for common or similar subunits. The actual biochemical role of the heterotrimeric complex in the damage response also remains unclear. While exonuclease activities have been reported for the counterparts of these proteins in other species, such activities have not been detected in budding yeast.

What is the role of the Rad24-RFC and Ddc1-Rad17-Mec3 complexes in the budding yeast damage response? Two separate studies in budding yeast found that the PCNA-like complex (Rad17/Mec3/Ddc1) was recruited to HO-induced DSBs, as well as to sites of *cdc13*-induced damage, in a Rad24-dependent manner.[5,6] This recruitment is independent of the Mec1/Ddc2 complex, suggesting that the two sets of proteins may respond independently to damage.[60] The human PCNA-like complex is similarly recruited onto chromatin in a Rad17-dependent, ATR-independent manner after UV irradiation. All of these proteins appear to become phosphorylated after damage, but the function/s of this damage-induced phosphorylation remain unresolved. Damage-induced phosphorylation is not required for formation of the PCNA-like complex in yeast or mammals, or for association of the proteins with chromatin at sites of damage,[6,49,61] although it may be associated with effector functions (e.g., ref. 62).

What is known, however, is that although these complexes translocate to sites of damage, they are not required for either initial recruitment of Mec1 to these sites, or for phosphorylation of at least a subset of proximal Mec1 targets.[63,64] The presence of PCNA-like complexes at damage sites could contribute to an elevated, or more targeted, level of Mec1 (PIKK) signalling, either by facilitating the processing of lesions to a structure or structures that are competent to send a checkpoint signal, or by somehow stabilising the interaction of Mec1 with lesions.

Recent data indicates that the PCNA-like and RFC-like complexes are recruited mainly to difficult-to-repair damage.[14,65-67] It is thought that much more damage occurs than is signalled by the damage checkpoint machinery in wild-type cells, as most of the DNA lesions that occur are rapidly repaired and do not trigger a checkpoint response.[65,68,69] If initial attempts at repair fail, lesion processing may generate intermediates that persist long enough to make productive interactions with checkpoint factors, resulting in the independent recruitment of PIKKs and PIKK substrates to damage sites. This is consistent with the observation that the human PIKK complex ATR (*A*taxia-*T*elangiectasia *R*elated)-ATRIP (*ATR-I*nteracting *P*rotein) can localise to damage sites independently of the RFC-like and PCNA-like complexes, but cannot activate a 'full' set of damage responses.[61] A requirement for multiple simultaneous activating signals might thus function as a 'fail-safe' mechanism, preventing inappropriate activation of the damage checkpoint. See Figure 2.

Mec1/Ddc2: Signal Transduction and More

In budding yeast, two proteins with a key role in transducing the DNA damage signal are Mec1 and Tel1, both members of the PIKK family of serine/threonine protein kinases that, unusually, display homology to a clan of lipid kinases. The PIKKs play central roles in a signalling network, activated in response to DNA damage, that is highly conserved from yeast to man. The Mec1 kinase is a key regulator of this signalling network in budding yeast. Its function is critical for cellular responses to many types of DNA damage and defects in the Mec1 pathway result in hypersensitivity to genotoxins, loss of DNA damage checkpoints and genomic instability.[25,70-72] Mec1 is a 270 kDa protein kinase with similarity in its catalytic domain to PI3K. Its kinase activity is required for all of its known functions.[73,74] Mec1 also has an essential role that is unrelated to its function in DNA damage checkpoints. *MEC1* deletion results in lethality which can be rescued by increasing cellular deoxyribonucleotide (dNTP) levels, either by overexpression of *RNR3*, a subunit of ribonucleotide reductase (RNR), or by deletion of *SML1*, an inhibitor of RNR.[75,76] However, elevation of cellular dNTP levels does not rescue the hypersensitivity to genotoxic agents or checkpoint defects in cells lacking functional Mec1.

In budding yeast, many types of DNA damage trigger a Mec1/Tel1-dependent protein phosphorylation cascade that ultimately impinges on the transcription, replication, repair and cell cycle machineries. Several cellular proteins, notably kinases such as Rad53[22,77,78] and Chk1,[32] become rapidly phosphorylated and activated in a Mec1/Tel1-dependent fashion. When activated, these kinases are able in turn to phosphorylate and modulate the activities of key effectors of the damage response. These include downstream targets involved in controlling cell cycle progression and repair, for example Dun1, Pds1, cyclin-dependent kinases (CDKs), CDK inhibitors and various repair proteins. Mec1 was originally identified by its involvement in transduction of the damage signal, although it has recently been linked more directly to the sensing process itself. Mec1 exists in a complex with Ddc2/Lcd1/Pie1 (hereafter referred to as Ddc2), that is required for all known functions of Mec1.[63,79,80] *DDC2* is required for all known DNA damage checkpoints and is the only known gene that, when deleted, shows the same severity of phenotype as a *MEC1* deletion. As with deletion of *MEC1*, the lethality resulting from *DDC2* disruption can be rescued by *RNR3* overexpression or *SML1* inactivation.

The Ddc2 protein has limited sequence similarity to fission yeast Rad26 ($Rad26^{Sp}$), which exists in a complex with the $Rad3^{Sp}$ kinase, the fission yeast homologue of Mec1. DNA damage-induced phosphorylation of $Rad26^{Sp}$ was found to depend on $Rad3^{Sp}$ but not on other known checkpoint proteins, suggesting that the Rad3-Rad26 complex might function at an early step in the damage recognition process.[4] Similarly to $Rad26^{Sp}$, *S. cerevisiae* Mec1 was subsequently shown to be essential for Ddc2 phosphorylation in response to damage, both in vitro and in vivo, but this phosphorylation does not require any of the other known checkpoint proteins, such as Ddc1, Rad53 or Pds1. Ddc2 is, however, required for DNA damage-induced phosphorylation of these proteins.[63] Thus the Ddc2/Mec1 complex appears to respond to DNA damage independently of other checkpoint proteins. This is also the case in human cells, where the Mec1/Rad3 counterpart ATR (*A*taxia-*T*elangiectasia *R*elated), functions in cooperation with ATR-interacting protein, ATRIP.[81] Ddc2 and its fission yeast and human counterparts $Rad26^{Sp}$ and ATRIP show poor sequence conservation, except in a region containing a conserved motif of four basic amino acids (KKRK). Mutations within this motif have no apparent effect on the stability or subcellular localisation of Ddc2, or its association with Mec1, but do abolish Ddc2 function, indicating that this motif is essential for function of the Ddc2/Mec1 complex.[80]

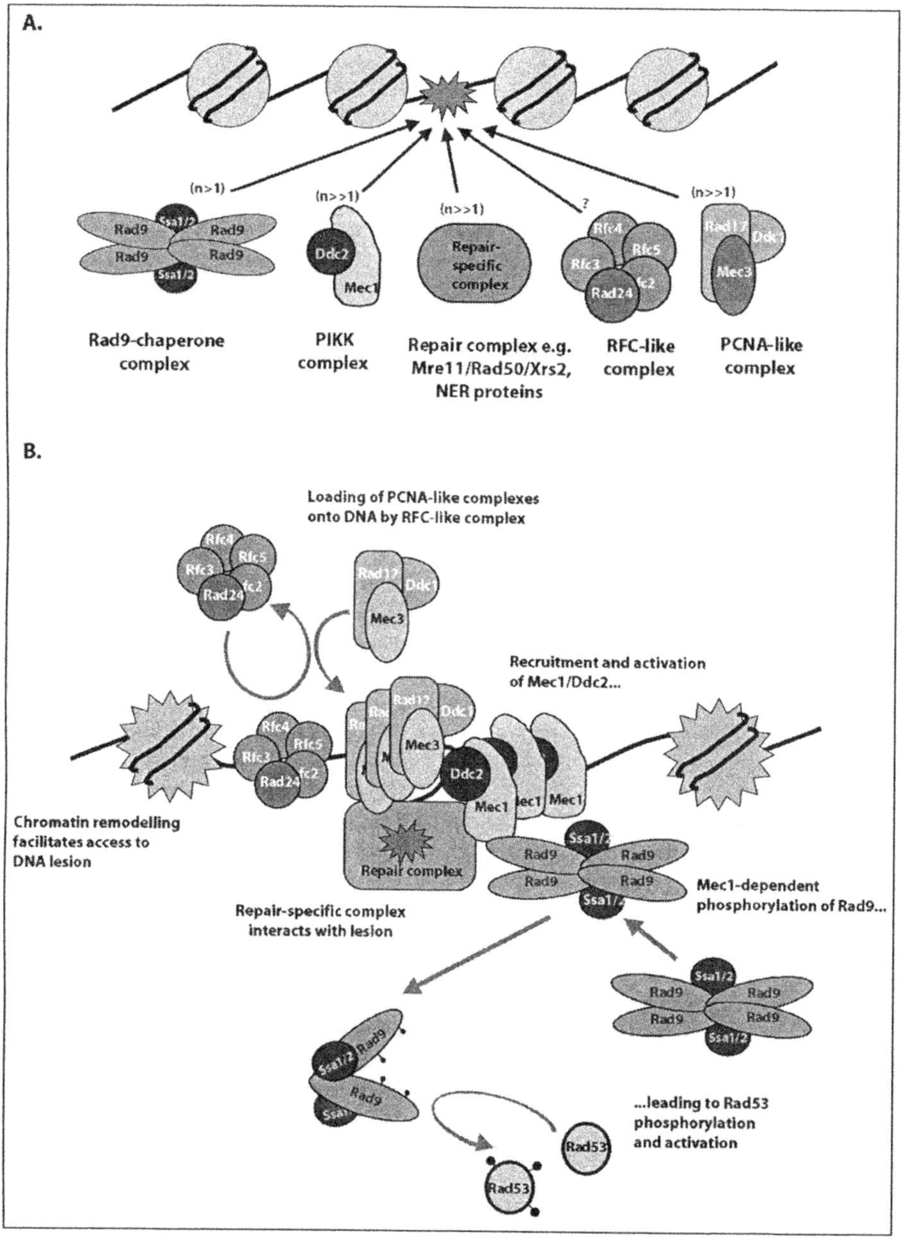

Figure 2. Recruitment of checkpoint and repair factors to lesions. A) Independent recruitment of different checkpoint complexes to lesions. The likely stoichiometry for each type of complex is indicated. A combination of independent activating signals may be required to trigger the global damage response—a 'fail-safe' mechanism (see text). B) After DNA damage, remodelling of chromatin surrounding the damage site facilitates access of specific repair complexes to the lesion, along with the Rad9/Ssa, PCNA-like and Mec1/Ddc2 complexes. For clarity, the Mec1/Ddc2 and PCNA-like complexes are shown binding at opposite sides of the lesion. These complexes can accumulate at persistent lesions, increasing the local concentration of damage signalling and repair activities, and promoting repair and/or amplification of the damage signal.

Can the Mecl/Ddc2 complex detect DNA damage directly? The Mecl/Ddc2 complex shows an ability to bind DNA in vitro and to localise to sites of damage in vivo.[5,6,60] The RFC-like and PCNA-like complexes have been proposed to function in the damage sensing process, by detecting damage, loading onto DNA and subsequently triggering Mecl activation. However, it has been shown that damage-stimulated phosphorylation of two of the most proximal Mecl targets, Ddc2 and the histone H2A, occurs independently of Rad9, Rad17 and Rad24.[63,64] Similarly, the fission yeast counterparts of these proteins are not required for $Rad3^{Sp}$-dependent phosphorylation of $Rad26^{Sp}$.[4] Recruitment to DNA lesions and initial activation of the Mecl/Ddc2 complex in response to *cdc13*-induced damage is not significantly affected by the absence of Rad9 or members of the *RAD24* epistasis group, although the activation of downstream effectors such as Rad53 is severely reduced.[60] Thus it appears that although sensing (and possibly processing) of damage by members of the *RAD24* epistasis group is necessary for full activation of downstream effectors, it is not required for initial Mecl recruitment or activation. Instead activation may occur directly, by interaction with structures associated with DNA damage (e.g., ssDNA), indirectly, through associated repair proteins, or through a combination of both mechanisms.

Zou et al have recently shown in human cells that the ssDNA-binding protein RPA (replication protein A) is required for ATR/ATRIP-dependent responses to DNA damage and replication stress.[81] Unlike ATM, ATR kinase activity is not stimulated by DNA damage, pointing to a different mode of regulation, most likely substrate recruitment. Binding of ATRIP to RPA-coated ssDNA enables the ATR/ATRIP complex to associate with DNA, via the interaction between ATRIP and RPA. ATR/ATRIP recruitment to ssDNA also stimulates phosphorylation of hRad17, one of its substrates on DNA. This appears to be a function conserved in eukaryotes, since these authors also showed that RPA is required for recruitment of Ddc2 (the budding yeast ATRIP counterpart) to DSBs in vivo. Checkpoint-defective yeast RPA mutants were found to be also defective in recruiting Ddc2 to HO-induced DSBs. Purified Ddc2 and its human equivalent, ATRIP, both showed a low affinity for DSBs in vivo relative to RPA-coated ssDNA. While this result does not rule out direct sensing of lesions (including DSBs) by Mecl/Ddc2 as previously reported,[60] it indicates a more significant role for RPA as a specificity factor, increasing the affinity of the ATR/ATRIP complex for ssDNA—a structure widely associated with damage/repair processing.

Reports of independent recruitment of the budding yeast Ddcl/Mec3/Rad17 and Mecl/Ddc2 complexes to DNA lesions in vivo[5,6,60] support a model in which an appropriate combination of activating signals is required to trigger a full-scale damage checkpoint response. An interesting speculation is that, once bound, the PCNA-like complexes might tether the originally relatively loosely bound Mecl/ATM/ATR (PIKKs) to the DNA and facilitate translocation of the PIKK in both directions, where it can then phosphorylate H2A/H2AX over megabase domains, thereby regulating the ordered assembly of the focal nuclear structures required for efficient signalling and repair. See Figure 3.

The Mre11 Complex: DSBs and Checkpoint Signalling

Of the various types of DNA lesion that occur in cells, double-strand breaks (DSBs) are probably the most dangerous. DSBs are generated when breaks in the two strands of the DNA double helix occur close together, so that base pairing and chromatin structure are insufficient to keep the DNA ends together. They are the major cytotoxic lesions caused by ionizing radiation or radiomimetic chemicals, but can also result from mechanical stress on chromosomes, or from a DNA (or possibly RNA) polymerase encountering nicks or single-strand breaks in the DNA. DSBs have potentially catastrophic biological consequences for the cell. In yeast, a single unrepairable double-strand break can be lethal[82-84] and in a metazoan cell, may be sufficient to

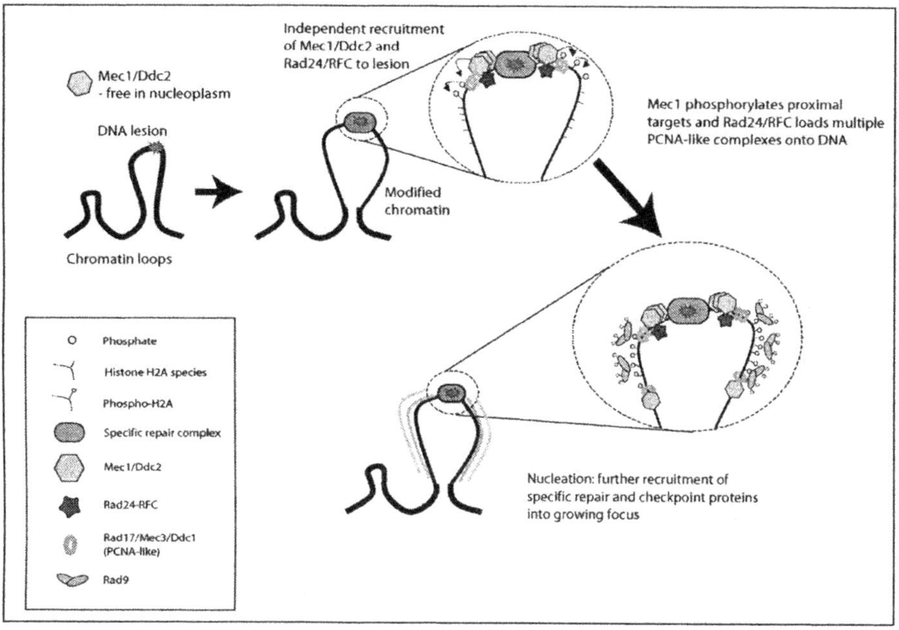

Figure 3. A model for focal recruitment of DNA damage response proteins. Modification of chromatin structure around a DNA lesion facilitates access of repair and checkpoint proteins to the damage site. Mec1/ Ddc2 and other checkpoint factors (e.g., Rad9 and the PCNA-like complex) independently recruited. Mec1 substrates around the lesion become phosphorylated; at a persistent lesion, interactions between multiple complexes help to recruit other checkpoint and specific repair proteins into a growing focus.

trigger apoptosis.[85] As might be expected for such a deleterious lesion, a single persistent DSB is sufficient to trigger G2/M arrest[10] and checkpoint-dependent redistribution of Ku and Sir proteins from telomeres to the break site.[86]

DSBs are intrinsically more difficult to repair than other types of damage. Firstly, the broken DNA ends are liable to become physically separated, preventing simple religation of the ends. Secondly, damage to the base and sugar components, as well as the phosphodiester backbone, particularly after ionizing radiation, frequently occurs at the DNA ends, generating complex breaks that are a further obstacle to rapid and error-free repair. Processing of the ends by many DNA metabolic enzymes, such as nucleotide kinases, nucleases and DNA polymerases, is often required before rejoining can occur. DSBs also pose a serious threat to genome integrity because of their potential for inducing genetic alteration; lack of, or inaccurate repair of, DSBs leads to genomic instability. DNA ends are highly recombinogenic and loss of heterozygosity, chromosomal loss and/or rearrangement can frequently be attributed to inappropriate recombination events.

The Mre11 complex consists of two highly conserved members, Mre11 and Rad50 and a third, relatively unconserved, member: Xrs2 in *S. cerevisiae* and Nbs1 in mammalian cells.[87] See reference 88 for a review of the mammalian Mre11 complex. In *S. cerevisiae*, the Mre11 complex has been implicated in recombinational DNA repair, nonhomologous end-joining (NHEJ), DSB end-resection, telomere maintenance and meiotic recombination (reviewed in ref. 15, 89). The Mre11 complex has also been implicated as a sensor of DSBs. In human cells, the hMre11/hRad50/Nbs1 complex rapidly associates with DSBs induced by IR.[90-92] Relocalisation of the complex is also observed during DNA replication in S phase.[15,93-95]

In budding yeast, the Mre11 complex is required for both of the major pathways, homologous recombination and nonhomologous end joining (NHEJ), for repair of double-strand breaks. Recent studies have indicated how the complex may function in tethering DNA ends[96,97] and in stimulating DNA end-joining by the DNA ligase Dnl4-Lif1.[98] Budding yeast Mre11 is an ~80 kDa protein that forms the core of the complex, interacting independently with both Rad50 and Xrs2.[99] Rad50 is a ~150 kDa protein that contains two ATP-binding motifs, at its amino and carboxyl terminal ends, that are essential for Rad50 function. The two ATP-binding domains are joined by an extended coiled-coil structure, interrupted by a putative globular domain, that is thought to function as a 'hinge', with the coiled-coil arm extending outwards from the core of the complex. Binding of ATP induces a conformational change that brings the two bipartite ATPase domains together in the presence of Mre11.[100]

Scanning force microscopy (SFM) of human Mre11/Rad50 complexes revealed specific association of multiple oligomers of these proteins with the ends of linear DNA molecules, via their globular domains.[96] Under some conditions, DNA molecules appeared to be tethered together by intermolecular interactions between the coiled-coil domains of Rad50. Oligomerisation of M/R complexes on DNA ends could create a kind of "molecular Velcro" to allow bridging of DNA ends. In *S. cerevisiae*, multiple linear DNA molecules tethered by large globular Mre11/Rad50/Xrs2 complexes have also been observed.[98] The end-bridging activity of the Mre11 complex may function to promote cellular survival of double-strand breaks, by holding the broken ends together, preventing loss of the broken chromatid fragment and also by stimulating end-joining by DNA ligases.

A role for the Mre11 complex in checkpoint signalling was first proposed based on observations of S phase checkpoint defects in cells from patients with Nijmegen breakage syndrome (NBS), a disease in which the third member of the human complex, Nbs1, is mutated. More recently, it has been demonstrated in budding yeast that in response to DSB-inducing agents, cells lacking a functional Mre11 complex are defective in the induction of G1 and G2 checkpoints, in addition to the intra-S checkpoint.[94] Yeast lacking a functional Mre11 complex are also defective in the HU-induced S-M checkpoint and the intra-S checkpoint induced by bleomycin treatment.[94,101] The S phase checkpoint defects are unlikely to be related to the role of the Mre11 complex in NHEJ or HR, since checkpoint control is unaffected by loss of these DNA repair pathways in a *rad52Δ yku70Δ* double mutant.[101] Work by Grenon et al extended the requirement for the Mre11 complex in the activation of DNA damage checkpoints to the G1 and G2 phases of the cell cycle. They demonstrated that an intact Mre11/Rad50/Xrs2 complex is required specifically in response to DSBs for the phosphorylation and activation of Rad9, Rad53 and Chk1, in the G1 and G2 phases of the cell cycle. Rad53 activation in G1- and G2-arrested *mre11Δ* cells is defective after gamma(γ)-irradiation or bleomycin treatment, but normal following ultraviolet (UV) irradiation. Cells harbouring null alleles of *RAD50* and *XRS2* behave similarly to *mre11Δ* cells, whereas mutation of *RAD51, RAD52* and *YKU80* did not result in detectable checkpoint defects. In addition, as would be expected for a protein complex that is believed to rapidly tether together broken chromosomes, analysis of checkpoint protein phosphorylation further suggests that the Mre11 complex functions at an early stage in the DSB response.[94]

The Mre11 complex is thus specifically required for efficient damage checkpoint activation in response to DSBs in all phases of the cell cycle. The complex is known to be capable of binding DSBs[91] and might therefore have an important role in sensing DSBs. Alternatively, it may be required for amplification of the checkpoint signal in response to a DSB, perhaps by regulating the generation of regions of ssDNA[17,83] that can stimulate checkpoint signalling by the ATR/ATM-like kinases. The available evidence is consistent with multiple roles for the Mre11 complex during the early phase of the DSB-specific checkpoint response. In addition to sensing, the complex may be required for amplification of the signal and also to act as a specificity factor for a subset of Mec1 substrates.

The potential role of this complex in recombinational repair of DNA DSBs has also been the subject of much debate. A number of recent studies indicate that most homologous recombination pathways are functional in Mre11 complex mutants and that this complex does not play a central role in the process of HR repair itself (reviewed in refs. 15, 102). The role of the Mre11 complex in DSB repair by NHEJ is likewise controversial; to date, clear evidence for its involvement is available only in budding yeast.[15] Perhaps the important tethering function that must occur very rapidly after the occurrence of a DSB is partially redundant with other activities. Alternatively, it is possible that the effect of Mre11 complex inactivation on DSB repair is due to defects in checkpoint function, suggesting that this complex could correspond to DNA damage-specific and checkpoint specific activity.

The S Phase Checkpoint

There are two principal checkpoint pathways that function in S phase in response to different types of stress: the DNA replication checkpoint and the intra-S checkpoint. Like the checkpoints in the G1 and G2 phases of the cell cycle, the S phase checkpoints coordinate cell cycle progression with DNA repair, prevent completion of S phase or premature entry to mitosis, stabilise stalled replication forks and allow DNA synthesis to be resumed when stress is removed or overcome.[19,103]

It is useful to consider the checkpoint response in S phase with respect to two distinct types of lesions, resulting from treatment with hydroxyurea (HU), an inhibitor of ribonucleotide reductase, an enzyme required for dNTP production, or with MMS (methyl-methanesulfonate), a DNA alkylating agent that causes single or double-strand breaks in S phase. HU-induced and MMS-induced lesions mainly trigger the "DNA replication" and the "intra-S" branches of the S phase checkpoint respectively (see below). Both branches activate checkpoint signalling through Mec1 and Rad53, but differ in the mechanism of Rad53 activation.

The DNA replication checkpoint, also known as the HU checkpoint, senses stresses at DNA replication forks that result when insufficient precursor nucleotides (dNTPs) are present or when the fork encounters a lesion on the DNA capable of blocking fork progression. This checkpoint is partially dependent on fork-associated replication proteins, such as DNA polymerase delta, its cofactor Dpb11, RFC and Sgs1p (reviewed in ref. 104). It is operationally defined as the checkpoint that causes cell cycle arrest and inhibits firing of late replication origins in response to DNA replication blocks.[105,106] Replication stress may be imposed by direct inhibition of DNA synthesis (e.g., by HU, an inhibitor of ribonucleotide reductase, an enzyme required for dNTP production) or may be associated with the replication machinery encountering DNA lesions, e.g., bleomycin, methyl-methanesulfonate.[20,107]

The "intra-S checkpoint" is the term used to describe that part of the DNA damage checkpoint response that is specific to S phase. Similarly to DNA damage checkpoints outside of S phase, roles for Rad9 and the Rad24 and Rad17 complexes have been described for the budding yeast intra-S phase checkpoint.[20,108-110] This checkpoint is believed to slow DNA replication by inhibiting Pol-α/primase activity[107] and possibly also by modifying the single-strand DNA binding protein, RPA.[111] The overall effect is a reduced rate of replication, due to a combination of slowed but stable fork progression and inhibition of late origin firing, allowing time for damage to be repaired or otherwise dealt with, until S phase has been completed.[103,112] Available evidence suggests at least two partially redundant branches of the pathway, one requiring Rad24, Rad17/Mec3/Ddc1 and Rad9, the other requiring Sgs1, a RecQ-family helicase.[20,110] Both branches act upstream of Rad53 to signal cell cycle arrest when DNA replication is perturbed.

The replication checkpoint and intra-S checkpoint pathways appear not only to act in parallel, but to cross-talk. MMS-induced DNA alkylation can stall replication forks (mimicking the effects of dNTP depletion) and HU-arrested forks are prone to strand breakage. Mrc1 is a highly charged, cell cycle regulated protein that may serve as the primary adaptor for checkpoint kinase activation in response to HU,[113,114] whereas Rad9 has a secondary role, helping to modulate Rad53p activation in S phase in response to MMS.

As the DNA damage checkpoint pathway is known to function in all phases of the cell cycle, it is possible that DNA damage that occurs in S phase can be directly sensed as damage, independently of the replication fork, in the same way as outside of S phase. However, it is clear that stresses imposed on elongating replication forks are an important mechanism for signalling damage in S phase. Replication stresses are believed to result in broken, stalled or collapsed replication forks, which may result in excessive ssDNA or a DSB at the replication fork. DSBs can also be generated when a replication fork passes through nicked DNA or certain repair/recombination intermediates,[115] or by agents such as topoisomerase I poisons.[69,116] Interestingly, recent data suggest that at least some types of topoisomerase I related lesions might be incapable of triggering efficient checkpoint activation, although Mec1-dependent phosphorylation of H2A is required for the repair of such lesions.[69]

Checkpoints and Genome Surveillance

Genetic studies in budding yeast first identified specific 'checkpoint' genes essential for cell cycle arrest after DNA damage. Since then, biochemical, physiological and structural studies in a wide range of eukaryotes (including budding and fission yeasts, as well as mice and humans) have added considerably to our understanding of damage checkpoint function and its relevance to cell behaviour. Recently, the emphasis has been on defining the biochemical mode of action of damage checkpoint components and a molecular picture of how eukaryotic cells sense and signal the presence of DNA damage is slowly beginning to emerge. The various damage checkpoints can be regarded as a single damage surveillance system, since diverse types of DNA damage are nevertheless signalled via common signal transduction pathways and trigger similar downstream responses.

What are the molecular signals that activate the DNA damage response? The presence of abnormal DNA structures, including single-stranded regions, double-strand breaks or DNA lesions themselves and aberrant replication fork structures, as well as alterations in higher-order chromatin structure, may all function as signals for damage checkpoint activation. The seemingly diverse modes of damage sensing employed by yeast and higher eukaryotes no doubt reflect this multiplicity of potential activating signals.

For example, human ATR appears to be able to directly bind to and become activated by UV photoproducts.[7] This is consistent with the possibility that, under certain circumstances, unprocessed primary lesions could contribute to the triggering of the damage checkpoint response. Alternatively, the DNA damage checkpoint may sense a particular type of lesion by interacting with repair proteins that can bind specifically to the lesion. Furthermore, lesion processing may be required for efficient checkpoint activation and/or amplification of the checkpoint signal.[14,58,67] Thus, under some conditions, lesion-specific repair factors may provide a molecular link between structures associated with damage and the checkpoint machinery.

The checkpoint response in NER-defective yeast cells may be due to conversion of UV-induced photoproducts into other structures (e.g., ssDNA, DSBs) that can be recognised independently of the NER proteins. This implies that any kind of lesion can be detected if it persists long enough to be converted to a form that can send a checkpoint signal. Consistent with this idea, not all DNA lesions are equally proficient at triggering what has been termed the 'global' damage response. The efficiency of the repair system involved in processing a particular type of lesion could play a large part in determining its ability to trigger checkpoints in the absence of DNA replication. A brief comparison of cellular responses to various types of damage tends to support this view. Endogenous oxidative damage, which is repaired by the base excision repair (BER) pathway, occurs extensively in aerobically metabolising cells, but does not normally trigger the damage checkpoint, presumable due to the efficiency of BER. In wild-type yeast, hydrogen peroxide-induced base damage does not result in Rad53 activation in G1 or G2, but does so in S phase. However, decreasing the efficiency of base excision repair by partial disruption of the BER pathway results in activation of Rad53-dependent checkpoints.[65] Similarly, Rad9-dependent G2/M checkpoint activation and cell death are triggered only when recognition of endogenous abasic lesions is prevented by deletion of *APN1/2* (BER) and *RAD1/RAD10* (NER) genes.[117] In contrast, for a highly deleterious lesion such as a double-strand break, the 'signalling threshold' can be very low indeed. A single persistent DSB can be detected and is sufficient to trigger relocalisation of Sir and Ku proteins to the site of the break.[10,86]

The observation that only persistent damage triggers the full checkpoint response has some interesting implications. Only unrepairable HO-induced DSBs trigger Rad53 activation[118] and the G2/M checkpoint is not activated if the break can be efficiently repaired by homologous recombination.[66] Thus lesion processing is not by itself sufficient to trigger checkpoint activation and it is likely that a large amount of 'silent' (with respect to checkpoint activation) repair occurs throughout the cell cycle. Observation of the dynamics of spontaneous Rad52 focus formation tends to support this view, as spontaneous Rad52 foci appear and disappear within minutes, without checkpoint activation.[68]

Focal Recruitment of Checkpoint and Repair Complexes to Damage Sites

DNA damage checkpoint pathways utilise a protein kinase cascade for signal amplification, resulting in an extraordinary sensitivity and rapidity of damage detection; a single double-strand break in the yeast genome (approximately 15 million base pairs, packaged into chromatin) can be sensed, triggering a rapid checkpoint response that includes Mec1- and Rad9-dependent G2/M arrest[83] and relocalisation of Sir proteins from telomeres to the break site.[86,119]

Many proteins involved in the damage response, including various checkpoint and repair components, become relocalised to sites of DNA damage in a manner similar to the Sir and Ku proteins. In mammalian cells, a large number of damage response proteins (including the Mre11 complex, BRCA1, ATR, ATRIP, Rad51 and Rad52) were found to form brightly-staining foci within the nucleus after treatment with DSB-inducing agents.[61,891,120-123] These nuclear foci almost certainly correspond to sites of DNA damage and repair. For example, X-ray irradiation of defined subnuclear volumes in human cells produces strong staining for Mre11/Rad50/Nbs1, a complex involved in various responses to DNA DSBs, only in the regions exposed to radiation.[91] In *S. cerevisiae* cells, a single HO-induced DSB gives rise to a single Ddc1 or Ddc2 focus, whereas *cdc13-1* induced telomeric damage gives rise to a number of foci that corresponds roughly to the number of telomere clusters present in the cell.[6]

The study of protein recruitment to DNA damage has been greatly facilitated by improved methods of generating discrete numbers and types of DNA lesions at precise sites within the genome. Such methods, in combination with chromatin immunoprecipitation (ChIP) and visualisation of proteins fused to GFP, have been successfully used by a number of groups to directly demonstrate the localisation of checkpoint complexes to sites of DNA damage. Two independent studies in budding yeast have shown that the Mec1/Ddc2 and the PCNA-like (Ddc1/Mec3/Rad17) complexes are recruited to damage sites independently of one another.[5,6] These observations have been confirmed in human cells, where association of the hRad9/hHus1/hRad1 complex with bulk chromatin was shown to require hRad17 (the human homologue of budding yeast Rad24), but not ATR.[61] See also Figure 2 for a model of how events occurring at a DNA lesion lead to damage checkpoint signalling.

Studies of the mobility of GFP-tagged proteins in living cells indicate that protein localisation to sites of DNA damage is a dynamic process. Using FRAP (Fluorescence recovery after photobleaching), it was shown in mammalian cells that IR-induced damage leads to transient immobilisation of the HR proteins Rad51, Rad52 and Rad54, consistent with their translocation to sites of DNA damage. In addition, the fluorescence of a single Rad52-containing focus was observed to recover over time, indicating that photobleached molecules within the focus could exchange with unbleached Rad52 molecules in the nucleoplasm.[123] Similar results were obtained with the NER repair complex ERCC1-XPF.[124]

It has now been directly demonstrated, using fluorescently-tagged proteins, that induced DSBs colocalise with foci containing Rad52, a DSB repair protein. Even after treatments sufficient to induce multiple DSBs, relatively few of these 'repair' foci (estimated to contain most of the Rad52 in the cell) are formed, which suggests that each focus is a site of repair for more than one lesion. Rad52 was also seen to form spontaneous foci in the absence of exogenous DNA damage. Time-lapse microscopy showed that the majority of these foci did not persist, but were rapidly assembled (most within ~30-90s) and disassembled (within about 10 min). A small fraction of cells contained persistent (>1 hr) Rad52 foci. These cells arrested and only resumed cycling after disappearance of the foci, suggesting that release from the checkpoint is coordinated with focus disassembly and presumably successful repair.[68]

Thus damage response factors can translocate to DNA lesions, but rather than becoming irreversibly bound, may exist in a state of dynamic equilibrium. Multiple damage recognition and repair complexes could then interact with a single DNA lesion within the same time-frame. In theory, this could provide opportunities for cross-talk or communication between various signalling pathways, or between different 'levels' in the same pathway. Such cross-talk might also account for the apparent lack of a linear order of function in the early phase of the damage response.

Large sub-nuclear structures such as DNA damage-induced foci are likely to be assembled around a scaffold that facilitates, organises and possibly even coordinates, transient and dynamic protein-DNA and protein-protein interactions. Such an architecture might allow sensors, transducers and repair factors to be brought into close proximity to each other, as well as to DNA lesions (see Fig. 3). This increase in the local concentration of repair and checkpoint factors may promote not only repair function, but also the efficient and rapid signalling of the status of DNA lesions within the cell.

The Chromatin Connection

Eukaryotic genomes are packaged within chromatin and it is within the context of chromatin that all DNA transactions occur. Chromatin structure modulates access of proteins to the DNA and is likely to affect other aspects of DNA processing, including DNA repair (reviewed in ref. 125). It is therefore plausible that DNA damage might result in altered chromatin structure and that these alterations might directly contribute to the checkpoint signal. Can the DNA damage checkpoint sense perturbations to chromatin structure that result from damage? To date, the evidence for this is largely circumstantial, although the converse, that DNA damage signalling and/or repair components can modulate chromatin structure, has certainly been demonstrated.

One of the earliest responses to DSB formation in yeast and vertebrates is phosphorylation of the C-terminal tail of histone H2A (H2AX in vertebrates) located in nucleosomes near the break. Downs et al showed that the Mec1-dependent phosphorylation of the C-terminal tail of the yeast core histone H2A at serine 129 in response to DNA damage is unrelated to Mec1-dependent transcriptional or cell cycle responses to damage.[64] This modification appears to result in alterations to chromatin structure suggested to contribute to efficient DSB repair by NHEJ; one possibility is that decondensation of chromatin surrounding the DNA lesion facilitates subsequent repair. A recent study has demonstrated an important role of the Mec1-dependent phosphorylation of budding yeast H2A in the efficient repair of camptothecin-induced DNA lesions that do not trigger a damage checkpoint response.[69] Thus this particular lesion (camptothecin causes capture of covalently linked topoisomerase I onto DNA) is 'checkpoint-blind' but still requires Mec1-dependent modification of chromatin for its repair.

In addition to facilitating the access or activity of repair factors, chromatin modifications might also have functions in signal transduction from lesions. Intriguingly, there is evidence to suggest that modification (in particular, acetylation) of histones can lead to activation of DNA damage checkpoints. Yeast carrying mutations in histone H4 that prevents its acetylation experience a transient G2/M delay that is dependent on the checkpoint protein Rad9.[41] More recently, studies have shown that deletion of components of the NuA4 histone acetylase (HAT) complex, which is specifically responsible for histone H4 acetylation, also result in a Rad9-dependent G2/M delay and damage sensitivity.[39,40] Chromatin assembly factors that package histones and newly replicated DNA into nucleosomes have also been implicated in the damage response. Asf1, a histone chaperone, deposits histone H3 and H4 onto newly synthesised DNA and yeast lacking Asf1 are sensitive to DSB-inducing agents.[126-128] *ASF1* deletion, HAT inactivation and the presence of nonacetylatable histone H4 would all result in a reduction of histone acetylation on chromatin, although how this might lead to activation of the damage checkpoint is not yet clear. It has been speculated that incomplete or inappropriate chromatin acetylation might be recognised as damage, thus triggering checkpoint activation.[40] Changes in histone acetylation are known to result in altered DNA supercoiling and one intriguing possibility is that the damage checkpoint might be capable of sensing altered DNA supercoiling resulting from DSBs, altered histone acetylation, or altered chromatin structure, in addition to events that occur at the damage site.

In human cells, a connection between chromatin and the damage checkpoint was initially suggested by the reported interaction between ATM and a histone deacetylase.[129] A recent landmark study on IR-induced ATM activation provides the strongest indication yet that DNA damage can be initially signalled by alterations in chromatin structure.[3] In mammalian cells, the ATM protein is crucial for initiating cellular responses to ionizing radiation (IR) and other DSB-inducing agents and is thought to act at an early stage of damage signal transduction. Bakkenist and Kastan found that IR-induced activation of ATM was extremely rapid and quickly reached maximal levels after exposure to even low-dose IR.[3] Furthermore, this modification was detectable after introduction of only a few strand breaks in the genome of a human cell, whether induced by low-dose IR or by a restriction enzyme. The rapidity of the modification and the large fraction of cellular ATM activated, relative to the low number of DSBs required to induce this activation, led the authors to conclude that ATM does not require direct binding to DSBs for its initial activation and to propose instead that strand breaks rapidly cause changes in higher-order chromatin structure that initiate activation of ATM at a distance from the DSB. They tested this hypothesis directly by treating cells with agents (e.g., trichostatin, chloroquine) known to perturb chromatin conformation without causing DNA damage and observed a diffuse nuclear phosphorylation of the ATM protein, in contrast to the formation of foci containing phosphorylated ATM and γ-H2AX seen after irradiation. Importantly, the chromatin-perturbing treatments used failed to induce phosphorylation of ATM substrates directly at break sites (e.g., H2AX), but could induce phosphorylation of known nucleoplasmic ATM substrates (e.g., p53) that are not localised to DSBs.[3] These data suggest that the protein kinase activity of ATM is somehow initially activated by the extremely rapid alteration to chromatin conformation caused by DSB, however, this activated form of ATM is nucleoplasmic and restricted to soluble nuclear targets. Subsequent recruitment of the activated ATM to DSB allows the already active, but now insoluble, kinase to target substrates directly at the break (e.g., H2AX) and those subsequently recruited to the break (e.g., the Mre11-Rad50-Nbs1 complex, BRCA1)

Concluding Remarks

The presence of a functional DNA checkpoint pathway is of critical importance to eukaryotic cells, primarily because of its role in maintaining genome stability. Checkpoint pathways in eukaryotes appear highly conserved and control not only cell cycle progression, but also other key aspects of cellular responses to DNA damage, such as DNA repair, recovery and/or adaptation. A key issue, that of precisely how the DNA damage signal is initiated, remains largely unresolved. Numerous proteins and protein complexes have been implicated in this process, but in the majority of cases, their precise function in damage detection has still to be fully elucidated. Nevertheless, a number of themes are emerging, with the identification of significant links between checkpoint function and genome stability in model organisms and in human cells. In many cases, DNA repair components appear to be the primary means for sensing a particular type of damage. Such components would bind to and perhaps process, the DNA lesions, but would also interact with checkpoint pathway components in order to initiate the damage signal. However, it is clear that the presence of damage can also be sensed in other, less direct, ways. This is unsurprising, given the myriad types of DNA lesions that can arise following genotoxic insult. It has long been thought that different kinds of lesions might be processed to a common structure, e.g., DSB, or single-stranded DNA, that can signal to checkpoint pathways. Signals arising at stalled or otherwise aberrant replication forks have also been hypothesised to be capable of triggering checkpoint activation, though their precise molecular nature remains open to speculation. A different and intriguing possibility, indicated by a recent study in mammalian cells, is that some forms of DNA damage might trigger changes, not only in the vicinity of the lesion (perhaps the chromosomal domain in which the lesion occurs), but that these changes can be somehow propagated throughout the nuclear compartment. This would suggest that future investigations should also consider long-range changes resulting from DNA damage and not just events occurring at sites of DNA damage.

References

1. Lindahl T. Instability and decay of the primary structure of DNA. Nature 1993; 362(6422):709-15.
2. Zhou BB, Elledge SJ. The DNA damage response: putting checkpoints in perspective. Nature 2000; 408(6811):433-9.
3. Bakkenist CJ, Kastan MB. DNA damage activates ATM through intermolecular autophosphorylation and dimer dissociation. Nature 2003; 421(6922):499-506.
4. Edwards RJ, Bentley NJ, Carr AM. A Rad3-Rad26 complex responds to DNA damage independently of other checkpoint proteins. Nat Cell Biol 1999; 1(7):393-398.
5. Kondo T, Wakayama T, Naiki T et al. Recruitment of Mec1 and Ddc1 checkpoint proteins to double-strand breaks through distinct mechanisms. Science 2001; 294(5543):867-70.
6. Melo JA, Cohen J, Toczyski DP. Two checkpoint complexes are independently recruited to sites of DNA damage in vivo. Genes Dev 2001; 15(21):2809-21.
7. Unsal-Kacmaz K, Makhov AM, Griffith JD et al. Preferential binding of ATR protein to UV-damaged DNA. Proc Natl Acad Sci USA 2002; 99(10):6673-8.
8. Toczyski DP, Galgoczy DJ, Hartwell LH. CDC5 and CKII control adaptation to the yeast DNA damage checkpoint. Cell 1997; 90(6):1097-106.
9. Vaze M, Pellicioli A, Lee S et al. Recovery from checkpoint-mediated arrest after repair of a double-strand break requires Srs2 helicase. Mol Cell 2002; 10(2):373.
10. Sandell LL, Zakian VA. Loss of a yeast telomere: arrest, recovery, and chromosome loss. Cell 1993; 75(4):729-39.

11. Sugasawa K, Ng JM, Masutani C et al. Xeroderma pigmentosum group C protein complex is the initiator of global genome nucleotide excision repair. Mol Cell 1998; 2(2):223-32.
12. Sugasawa K, Okamoto T, Shimizu Y et al. A multistep damage recognition mechanism for global genomic nucleotide excision repair. Genes Dev 2001; 15(5):507-21.
13. Hey T, Lipps G, Sugasawa K et al. The XPC-HR23B complex displays high affinity and specificity for damaged DNA in a true-equilibrium fluorescence assay. Biochemistry 2002; 41(21):6583-7.
14. Neecke H, Lucchini G, Longhese MP. Cell cycle progression in the presence of irreparable DNA damage is controlled by a Mec1- and Rad53-dependent checkpoint in budding yeast. Embo J 1999; 18(16):4485-97.
15. D'Amours D, Jackson SP. The Mre11 complex: at the crossroads of DNA repair and checkpoint signalling. Nat Rev Mol Cell Biol 2002; 3(5):317-27.
16. Lydall D, Nikolsky Y, Bishop DK et al. A meiotic recombination checkpoint controlled by mitotic checkpoint genes. Nature 1996; 383(6603):840-3.
17. Lydall D, Weinert T. Yeast checkpoint genes in DNA damage processing: implications for repair and arrest. Science 1995; 270(5241):1488-91.
18. Garvik B, Carson M, Hartwell L. Single-stranded DNA arising at telomeres in cdc13 mutants may constitute a specific signal for the RAD9 checkpoint. Mol Cell Biol 1995; 15(11):6128-38.
19. Osborn AJ, Elledge SJ, Zou L. Checking on the fork: the DNA-replication stress-response pathway. Trends Cell Biol 2002; 12(11):509-16.
20. Myung K, Kolodner RD. Suppression of genome instability by redundant S-phase checkpoint pathways in Saccharomyces cerevisiae. Proc Natl Acad Sci USA 2002; 99(7):4500-7.
21. Hartwell LH, Weinert TA. Checkpoints: controls that ensure the order of cell cycle events. Science 1989; 246(4930):629-34.
22. de la Torre-Ruiz MA, Green CM, Lowndes NF. RAD9 and RAD24 define two additive, interacting branches of the DNA damage checkpoint pathway in budding yeast normally required for Rad53 modification and activation. EMBO J 1998; 17(9):2687-98.
23. Weinert TA, Hartwell LH. The RAD9 gene controls the cell cycle response to DNA damage in Saccharomyces cerevisiae. Science 1988; 241(4863):317-22.
24. Siede W, Friedberg AS, Friedberg EC. RAD9-dependent G1 arrest defines a second checkpoint for damaged DNA in the cell cycle of Saccharomyces cerevisiae. Proc Natl Acad Sci USA 1993; 90(17):7985-9.
25. Paulovich AG, Hartwell LH. A checkpoint regulates the rate of progression through S phase in S. cerevisiae in response to DNA damage. Cell 1995; 82:841-847.
26. Weinert TA, Hartwell LH. Characterization of RAD9 of Saccharomyces cerevisiae and evidence that its function acts posttranslationally in cell cycle arrest after DNA damage. Mol Cell Biol 1990; 10(12):6554-64.
27. Watanabe K, Morishita J, Umezu K et al. Involvement of RAD9-dependent damage checkpoint control in arrest of cell cycle, induction of cell death, and chromosome instability caused by defects in origin recognition complex in Saccharomyces cerevisiae. Eukaryot Cell 2002; 1(2):200-12.
28. Emili A. MEC1-dependent phosphorylation of Rad9p in response to DNA damage. Mol Cell 1998; 2(2):183-9.
29. Soulier J, Lowndes NF. The BRCT domain of the S. cerevisiae checkpoint protein Rad9 mediates a Rad9-Rad9 interaction after DNA damage. Curr Biol 1999; 9(10):551-4.
30. Sun Z, Hsiao J, Fay DS et al. Rad53 FHA domain associated with phosphorylated Rad9 in the DNA damage checkpoint. Science 1998; 281(5374):272-4.
31. Vialard JE, Gilbert CS, Green CM et al. The budding yeast Rad9 checkpoint protein is subjected to Mec1/Tel1-dependent hyperphosphorylation and interacts with Rad53 after DNA damage. Embo J 1998; 17(19):5679-88.
32. Sanchez Y, Bachant J, Wang H et al. Control of the DNA damage checkpoint by chk1 and rad53 protein kinases through distinct mechanisms. Science 1999; 286(5442):1166-1171.
33. Navas TA, Sanchez Y, Elledge SJ. RAD9 and DNA polymerase epsilon form parallel sensory branches for transducing the DNA damage checkpoint signal in Saccharomyces cerevisiae. Genes Dev 1996; 10(20):2632-43.
34. Schwartz MF, Duong JK, Sun Z et al. Rad9 phosphorylation sites couple Rad53 to the Saccharomyces cerevisiae DNA damage checkpoint. Mol Cell 2002; 9(5):1055-65.

35. Gilbert CS, Green CM, Lowndes NF. Budding yeast Rad9 is an ATP-dependent Rad53 activating machine. Mol Cell 2001; 8(1):129-36.
36. Durocher D, Henckel J, Fersht AR et al. The FHA domain is a modular phosphopeptide recognition motif. Mol Cell 1999; 4(3):387-94.
37. Gilbert CS, van den Bosch M, Green CM et al. The budding yeast Rad9 checkpoint complex: chaperone proteins are required for its function. EMBO Rep 2003; 4(10):953-958.
38. Weinert T. DNA damage checkpoints update: getting molecular. Curr Opin Genet Dev 1998; 8(2):185-93.
39. Choy JS, Tobe BT, Huh JH et al. Yng2p-dependent NuA4 histone H4 acetylation activity is required for mitotic and meiotic progression. J Biol Chem 2001; 276(47):43653-62.
40. Clarke AS, Lowell JE, Jacobson SJ et al. Esa1p is an essential histone acetyltransferase required for cell cycle progression. Mol Cell Biol 1999; 19(4):2515-26.
41. Megee PC, Morgan BA, Smith MM. Histone H4 and the maintenance of genome integrity. Genes Dev 1995; 9(14):1716-27.
42. Griffiths DJ, Barbet NC, McCready S et al. Fission yeast rad17: a homologue of budding yeast RAD24 that shares regions of sequence similarity with DNA polymerase accessory proteins. Embo J 1995; 14(23):5812-23.
43. Thelen MP, Venclovas C, Fidelis K. A sliding clamp model for the Rad1 family of cell cycle checkpoint proteins. Cell 1999; 96(6):769-70.
44. Aravind L, Walker DR, Koonin EV. Conserved domains in DNA repair proteins and evolution of repair systems. Nucleic Acids Res 1999; 27(5):1223-42.
45. Mossi R, Hubscher U. Clamping down on clamps and clamp loaders—the eukaryotic replication factor C. Eur J Biochem 1998; 254(2):209-16.
46. Shimomura T, Ando S, Matsumoto K et al. Functional and physical interaction between Rad24 and Rfc5 in the yeast checkpoint pathways. Mol Cell Biol 1998; 18(9):5485-91.
47. Green CM, Erdjument-Bromage H, Tempst P et al. A novel Rad24 checkpoint protein complex closely related to replication factor C. Curr Biol 2000; 10(1):39-42.
48. Venclovas C, Thelen MP. Structure-based predictions of Rad1, Rad9, Hus1 and Rad17 participation in sliding clamp and clamp-loading complexes. Nucleic Acids Res 2000; 28(13):2481-93.
49. Kondo T, Matsumoto K, Sugimoto K. Role of a complex containing Rad17, Mec3, and Ddc1 in the yeast DNA damage checkpoint pathway. Mol Cell Biol 1999; 19(2):1136-43.
50. Shiomi Y, Shinozaki A, Nakada D et al. Clamp and clamp loader structures of the human checkpoint protein complexes, Rad9-1-1 and Rad17-RFC. Genes Cells 2002; 7(8):861-8.
51. Lindsey-Boltz LA, Bermudez VP, Hurwitz J et al. Purification and characterization of human DNA damage checkpoint Rad complexes. Proc Natl Acad Sci USA 2001; 98(20):11236-41.
52. Burtelow MA, Roos-Mattjus PM, Rauen M et al. Reconstitution and molecular analysis of the hRad9-hHus1-hRad1 (9-1-1) DNA damage responsive checkpoint complex. J Biol Chem 2001; 276(28):25903-25909.
53. Majka J, Burgers PM. Yeast Rad17/Mec3/Ddc1: A sliding clamp for the DNA damage checkpoint. Proc Natl Acad Sci USA 2003; 100(5):2249-54.
54. Bermudez VP, Lindsey-Boltz LA, Cesare AJ et al. Loading of the human 9-1-1 checkpoint complex onto DNA by the checkpoint clamp loader hRad17-replication factor C complex in vitro. Proc Natl Acad Sci USA 2003; 100(4):1633-8.
55. Naiki T, Kondo T, Nakada D et al. Chl12 (Ctf18) forms a novel replication factor C-related complex and functions redundantly with Rad24 in the DNA replication checkpoint pathway. Mol Cell Biol 2001; 21(17):5838-45.
56. Mayer ML, Gygi SP, Aebersold R et al. Identification of RFC(Ctf18p, Ctf8p, Dcc1p): an alternative RFC complex required for sister chromatid cohesion in S. cerevisiae. Mol Cell 2001; 7(5):959-70.
57. Kenna MA, Skibbens RV. Mechanical link between cohesion establishment and DNA replication: Ctf7p/Eco1p, a cohesion establishment factor, associates with three different replication factor C complexes. Mol Cell Biol 2003; 23(8):2999-3007.
58. Paulovich AG, Armour CD, Hartwell LH. The Saccharomyces cerevisiae RAD9, RAD17, RAD24 and MEC3 genes are required for tolerating irreparable, ultraviolet-induced DNA damage. Genetics 1998; 150(1):75-93.

59. Zhang H, Zhu Z, Vidanes G et al. Characterization of DNA damage-stimulated self-interaction of Saccharomyces cerevisiae checkpoint protein Rad17p. J Biol Chem 2001; 276(28):26715-23.
60. Rouse J, Jackson SP. Lcd1p recruits Mec1p to DNA lesions in vitro and in vivo. Mol Cell 2002; 9(4):857-69.
61. Zou L, Cortez D, Elledge SJ. Regulation of ATR substrate selection by Rad17-dependent loading of Rad9 complexes onto chromatin. Genes Dev 2002; 16(2):198-208.
62. Roos-Mattjus P, Hopkins KM, Oestreich AJ et al. Phosphorylation of human Rad9 is required for genotoxin-activated checkpoint signaling. J Biol Chem 2003.
63. Paciotti V, Clerici M, Lucchini G et al. The checkpoint protein Ddc2, functionally related to S. pombe Rad26, interacts with Mec1 and is regulated by Mec1-dependent phosphorylation in budding yeast. Genes Dev 2000; 14(16):2046-59.
64. Downs JA, Lowndes NF, Jackson SP. A role for Saccharomyces cerevisiae histone H2A in DNA repair. Nature 2000; 408(6815):1001-4.
65. Leroy C, Mann C, Marsolier MC. Silent repair accounts for cell cycle specificity in the signaling of oxidative DNA lesions. Embo J 2001; 20(11):2896-906.
66. Pellicioli A, Lee SE, Lucca C et al. Regulation of Saccharomyces Rad53 checkpoint kinase during adaptation from DNA damage-induced G2/M arrest. Mol Cell 2001; 7(2):293-300.
67. Freire R, Murguia JR, Tarsounas M et al. Human and mouse homologs of Schizosaccharomyces pombe rad1(+) and Saccharomyces cerevisiae RAD17: linkage to checkpoint control and mammalian meiosis. Genes Dev 1998; 12(16):2560-73.
68. Lisby M, Mortensen UH, Rothstein R. Colocalization of multiple DNA double-strand breaks at a single Rad52 repair centre. Nat Cell Biol 2003; 5(6):572-7.
69. Redon C, Pilch DR, Rogakou EP et al. Yeast histone 2A serine 129 is essential for the efficient repair of checkpoint-blind DNA damage. EMBO Rep 2003; 4(7):1-7.
70. Desany BA, Alcasabas AA, Bachant JB et al. Recovery from DNA replicational stress is the essential function of the S-phase checkpoint pathway. Genes Dev 1998; 12(18):2956-70.
71. Sanchez Y, Desany BA, Jones WJ et al. Regulation of RAD53 by the ATM-like kinases MEC1 and TEL1 in yeast cell cycle checkpoint pathways. Science 1996; 271(5247):357-60.
72. Weinert TA, Kiser GL, Hartwell LH. Mitotic checkpoint genes in budding yeast and the dependence of mitosis on DNA replication and repair. Genes Dev 1994; 8(6):652-65.
73. Paciotti V, Clerici M, Scotti M et al. Characterization of mec1 kinase-deficient mutants and of new hypomorphic mec1 alleles impairing subsets of the DNA damage response pathway. Mol Cell Biol 2001; 21(12):3913-25.
74. Mallory JC, Petes TD. Protein kinase activity of Tel1p and Mec1p, two Saccharomyces cerevisiae proteins related to the human ATM protein kinase. Proc Natl Acad Sci USA 2000; 97(25):13749-54.
75. Zhao X, Georgieva B, Chabes A et al. Mutational and structural analyses of the ribonucleotide reductase inhibitor Sml1 define its Rnr1 interaction domain whose inactivation allows suppression of mec1 and rad53 lethality. Mol Cell Biol 2000; 20(23):9076-83.
76. Zhao X, Muller EG, Rothstein R. A suppressor of two essential checkpoint genes identifies a novel protein that negatively affects dNTP pools. Mol Cell 1998; 2(3):329-40.
77. Allen JB, Zhou Z, Siede W et al. The SAD1/RAD53 protein kinase controls multiple checkpoints and DNA damage-induced transcription in yeast. Genes Dev 1994; 8:2401-15.
78. Sun Z, Fay DS, Marini F et al. Spk1/Rad53 is regulated by Mec1-dependent protein phosphorylation in DNA replication and damage checkpoint pathways. Genes Dev 1996; 10(4):395-406.
79. Rouse J, Jackson SP. LCD1: an essential gene involved in checkpoint control and regulation of the MEC1 signalling pathway in Saccharomyces cerevisiae. Embo J 2000; 19(21):5801-12.
80. Wakayama T, Kondo T, Ando S et al. Pie1, a protein interacting with Mec1, controls cell growth and checkpoint responses in Saccharomyces cerevisiae. Mol Cell Biol 2001; 21(3):755-64.
81. Zou L, Elledge SJ. Sensing DNA damage through ATRIP recognition of RPA-ssDNA complexes. Science 2003; 300(5625):1542-8.
82. Bennett CB, Lewis AL, Baldwin KK et al. Lethality induced by a single site-specific double-strand break in a dispensable yeast plasmid. Proc Natl Acad Sci USA 1993; 90(12):5613-7.
83. Lee SE, Moore JK, Holmes A et al. Saccharomyces Ku70, mre11/rad50 and RPA proteins regulate adaptation to G2/M arrest after DNA damage. Cell 1998; 94(3):399-409.

84. Lewis LK, Resnick MA. Tying up loose ends: nonhomologous end-joining in Saccharomyces cerevisiae. Mutat Res 2000; 451(1-2):71-89.
85. Huang LC, Clarkin KC, Wahl GM. Sensitivity and selectivity of the DNA damage sensor responsible for activating p53-dependent G1 arrest. Proc Natl Acad Sci USA 1996; 93(10):4827-32.
86. Mills KD, Sinclair DA, Guarente L. MEC1-dependent redistribution of the Sir3 silencing protein from telomeres to DNA double-strand breaks. Cell 1999; 97(5):609-20.
87. Maser RS, Mirzoeva OK, Wells J et al. Mre11 complex and DNA replication: linkage to E2F and sites of DNA synthesis. Mol Cell Biol 2001; 21(17):6006-16.
88. van den Bosch M, Bree RT, Lowndes NF. The MRN complex: coordinating and mediating the response to broken chromosomes. EMBO Rep 2003; 4(9):844-849.
89. Haber JE. The many interfaces of Mre11. Cell 1998; 95(5):583-6.
90. Mirzoeva OK, Petrini JH. DNA damage-dependent nuclear dynamics of the Mre11 complex. Mol Cell Biol 2001; 21(1):281-8.
91. Nelms BE, Maser RS, MacKay JF et al. In situ visualization of DNA double-strand break repair in human fibroblasts. Science 1998; 280(5363):590-2.
92. Carney JP, Maser RS, Olivares H et al. The hMre11/hRad50 protein complex and Nijmegen breakage syndrome: linkage of double-strand break repair to the cellular DNA damage response. Cell 1998; 93(3):477-86.
93. Costanzo V, Robertson K, Bibikova M et al. Mre11 protein complex prevents double-strand break accumulation during chromosomal DNA replication. Mol Cell 2001; 8:137–147.
94. Grenon M, Gilbert C, Lowndes NF. Checkpoint activation in response to double-strand breaks requires the Mre11/Rad50/Xrs2 complex. Nat Cell Biol 2001; 3(9):844-7.
95. Usui T, Ogawa H, Petrini JH. A DNA damage response pathway controlled by Tel1 and the Mre11 complex. Mol Cell 2001; 7(6):1255-66.
96. de Jager M, van Noort J, van Gent DC et al. Human Rad50/Mre11 is a flexible complex that can tether DNA ends. Mol Cell 2001; 8(5):1129-35.
97. Hopfner KP, Craig L, Moncalian G et al. The Rad50 zinc-hook is a structure joining Mre11 complexes in DNA recombination and repair. Nature 2002; 418(6897):562-6.
98. Chen L, Trujillo K, Ramos W et al. Promotion of Dnl4-catalyzed DNA end-joining by the Rad50/Mre11/Xrs2 and Hdf1/Hdf2 complexes. Mol Cell 2001; 8(5):1105-15.
99. Usui T, Ohta T, Oshiumi H et al. Complex formation and functional versatility of Mre11 of budding yeast in recombination. Cell 1998; 95(5):705-16.
100. Hopfner KP, Karcher A, Craig L et al. Structural biochemistry and interaction architecture of the DNA double-strand break repair Mre11 nuclease and Rad50-ATPase. Cell 2001; 105(4):473-85.
101. D'Amours D, Jackson SP. The yeast Xrs2 complex functions in S phase checkpoint regulation. Genes Dev 2001; 15(17):2238-49.
102. Paull TT. New glimpses of an old machine. Cell 2001; 107(5):563-5.
103. Kolodner RD, Putnam CD, Myung K. Maintenance of genome stability in Saccharomyces cerevisiae. Science 2002; 297(5581):552-7.
104. Foiani M, Pellicioli A, Lopes M et al. DNA damage checkpoints and DNA replication controls in Saccharomyces cerevisiae. Mutat Res 2000; 451(1-2):187-96.
105. Santocanale C, Diffley JF. A Mec1- and Rad53-dependent checkpoint controls late-firing origins of DNA replication. Nature 1998; 395(6702):615-8.
106. Shirahige K, Hori Y, Shiraishi K et al. Regulation of DNA-replication origins during cell-cycle progression. Nature 1998; 395:618-621.
107. Marini F, Pellicioli A, Paciotti V et al. A role for DNA primase in coupling DNA replication to DNA damage response. Embo J 1997; 16(3):639-50.
108. Lydall D, Weinert T. G2/M checkpoint genes of Saccharomyces cerevisiae: further evidence for roles in DNA replication and/or repair. Mol Gen Genet 1997; 256(6):638-51.
109. Paulovich AG, Margulies RU, Garvik BM et al. RAD9, RAD17, and RAD24 are required for S phase regulation in Saccharomyces cerevisiae in response to DNA damage. Genetics 1997; 145(1):45-62.
110. Frei C, Gasser SM. The yeast Sgs1p helicase acts upstream of Rad53p in the DNA replication checkpoint and colocalizes with Rad53p in S-phase-specific foci. Genes Dev 2000; 14(1):81-96.

111. Longhese MP, Neecke H, Paciotti V et al. The 70 kDa subunit of replication protein A is required for the G1/S and intra-S DNA damage checkpoints in budding yeast. Nucleic Acids Res 1996; 24(18):3533-7.
112. Tercero JA, Diffley JF. Regulation of DNA replication fork progression through damaged DNA by the Mecl/Rad53 checkpoint. Nature 2001; 412(6846):553-7.
113. Alcasabas AA, Osborn AJ, Bachant J et al. Mrcl transduces signals of DNA replication stress to activate Rad53. Nat Cell Biol 2001; 3(11):958-65.
114. Tanaka K, Russell P. Mrcl channels the DNA replication arrest signal to checkpoint kinase Cds1. Nat Cell Biol 2001; 3(11):966-72.
115. Cliby WA, Roberts CJ, Cimprich KA et al. Overexpression of a kinase-inactive ATR protein causes sensitivity to DNA-damaging agents and defects in cell cycle checkpoints. Embo J 1998; 17(1):159-69.
116. Furuta T, Takemura H, Liao ZY et al. Phosphorylation of histone H2AX and activation of Mre11, Rad50, and Nbs1 in response to replication-dependent DNA-double-strand breaks induced by mammalian DNA topoisomerase I cleavage complexes. J Biol Chem 2003.
117. Guillet M, Boiteux S. Endogenous DNA abasic sites cause cell death in the absence of Apn1, Apn2 and Rad1/Rad10 in Saccharomyces cerevisiae. Embo J 2002; 21(11):2833-41.
118. Lee SE, Paques F, Sylvan J et al. Role of yeast SIR genes and mating type in directing DNA double-strand breaks to homologous and non-homologous repair paths. Curr Biol 1999; 9(14):767-70.
119. Martin SG, Laroche T, Suka N et al. Relocalization of telomeric Ku and SIR proteins in response to DNA strand breaks in yeast. Cell 1999; 97(5):621-33.
120. Scully R, Chen J, Ochs RL et al. Dynamic changes of BRCA1 subnuclear location and phosphorylation state are initiated by DNA damage. Cell 1997; 90(3):425-35.
121. Maser RS, Monsen KJ, Nelms BE et al. hMre11 and hRad50 nuclear foci are induced during the normal cellular response to DNA double-strand breaks. Mol Cell Biol 1997; 17(10):6087-96.
122. Haaf T, Golub EI, Reddy G et al. Nuclear foci of mammalian Rad51 recombination protein in somatic cells after DNA damage and its localization in synaptonemal complexes. Proc Natl Acad Sci USA 1995; 92(6):2298-302.
123. Essers J, Houtsmuller AB, van Veelen L et all. Nuclear dynamics of RAD52 group homologous recombination proteins in response to DNA damage. Embo J 2002; 21(8):2030-7.
124. Houtsmuller AB, Rademakers S, Nigg AL et al. Action of DNA repair endonuclease ERCC1/XPF in living cells. Science 1999; 284(5416):958-61.
125. Green CM, Almouzni G. When repair meets chromatin. First in series on chromatin dynamics. EMBO Rep 2002; 3(1):28-33.
126. Bennett CB, Snipe JR, Westmoreland JW et al. SIR functions are required for the toleration of an unrepaired double-strand break in a dispensable yeast chromosome. Mol Cell Biol 2001; 21(16):5359-73.
127. Mello JA, Sillje HH, Roche DM et al. Human Asf1 and CAF-1 interact and synergize in a repair-coupled nucleosome assembly pathway. EMBO Rep 2002; 3(4):329-34.
128. Tyler JK, Adams CR, Chen SR et al. The RCAF complex mediates chromatin assembly during DNA replication and repair. Nature 1999; 402(6761):555-60.
129. Kim GD, Choi YH, Dimtchev A et al. Sensing of ionizing radiation-induced DNA damage by ATM through interaction with histone deacetylase. J Biol Chem 1999; 274(44):31127-30.
130. Sugimoto K, Ando S, Shimomura T, Matsumoto K. Rfc5, a replication factor C component, is required for regulation of Rad53 protein kinase in the yeast checkpoint pathway. Mol Cell Biol 1997; 17(10):5905-14.
131. Naiki T, Shimomura T, Kondo T et al. Rfc5, in cooperation with Rad24, controls DNA damage checkpoints throughout the cell cycle in Saccharomyces cerevisiae. Mol Cell Biol 2000; 20(16):5888-96.
132. Paciotti V, Lucchini G, Plevani P et al. Mec1p is essential for phosphorylation of the yeast DNA damage checkpoint protein Ddc1p, which physically interacts with Mec3p. EMBO J 1998; 17(14):4199-209.
133. Weinert T, Hartwell L. Control of G2 delay by the rad9 gene of Saccharomyces cerevisiae. J Cell Sci Suppl 1989; 12:145-8.

134. Aboussekhra A, Vialard JE, Morrison DE et al. A novel role for the budding yeast RAD9 checkpoint gene in DNA damage-dependent transcription. EMBO J 1996; 15(15):3912-22.
135. Araki H, Leem SH, Phongdara A et al. Dpb11, which interacts with DNA polymerase II(epsilon) in Saccharomyces cerevisiae, has a dual role in S-phase progression and at a cell cycle checkpoint. Proc Natl Acad Sci USA 1995; 92(25):11791-5.
136. Wang H, Elledge SJ. DRC1, DNA replication and checkpoint protein 1, functions with DPB11 to control DNA replication and the S-phase checkpoint in Saccharomyces cerevisiae. Proc Natl Acad Sci USA 1999; 96(7):3824-9.
137. Wang H, Elledge SJ. Genetic and physical interactions between DPB11 and DDC1 in the yeast DNA damage response pathway. Genetics 2002; 160(4):1295-304.
138. Liberi G, Chiolo I, Pellicioli A et al. Srs2 DNA helicase is involved in checkpoint response and its regulation requires a functional Mec1-dependent pathway and Cdk1 activity. Embo J 2000; 19(18):5027-38.
139. Lopes M, Cotta-Ramusino C, Pellicioli A et al. The DNA replication checkpoint response stabilizes stalled replication forks. Nature 2001; 412(6846):557-61.

Chapter 2

Mammalian Base Excision Repair

Grigory L. Dianov, Sarah L. Allinson, Helen Budworth and Kate Sleeth

Abstract

In living cells DNA base lesions are formed continuously as a consequence of normal metabolism and are also generated by a number of external factors. Simple base damages are repaired by base excision repair that includes two major pathways. These two pathways involve different subsets of enzymes and result in replacement of one (short-patch pathway) or more (long-patch pathway) nucleotides. Both pathways are initiated by a damage specific DNA glycosylase, which removes the damaged base creating an abasic site (apurinic/apyrimidinic, AP site). AP endonuclease then cleaves the phosphodiester bond 5' to the AP site and then DNA polymerase β adds the first nucleotide to the 3'-end of the incised AP site. Normally, the reaction continues through the short-patch repair pathway where Pol β removes the 5'-sugar phosphate residue and DNA ligase complete the repair. However, if the 5'-sugar phosphate is resistant to β-elimination, catalyzed by DNA polymerase β, then additional DNA synthesis is required to displace the 5'-sugar phosphate as part of a flap. This flap is then removed by flap endonuclease, and DNA ligase completes long-patch repair by ligating the DNA ends. These processes are directed and coordinated by multiple protein-protein interactions.

Introduction

DNA is an inherently unstable molecule and is subject to chemical degradation through processes such as hydrolysis, oxidation and alkylation.[1] This instability of the primary structure of DNA may lead to a loss or corruption of the information held within the genome with potentially deleterious effects for the organism. The aqueous environment of the cell is itself a source of endogenous DNA damage, with many of the chemical bonds of DNA being labile under physiological conditions. For example, it has been estimated that the instability of the N-glycosylic bond of purines to hydrolysis may result in the formation of abasic (AP) sites at a rate of approximately 10, 000 per human genome per day.[2] The accumulation of unrepaired AP sites can block DNA synthesis and cause cell death and mutations.[3,4]

In addition to the lability of the N-glycosylic bond, the exocyclic amino groups of the DNA bases are subject to hydrolytic deamination.[5] In the case of cytosine, this results in the formation of uracil which, if not repaired, can directly mispair with adenine, generating C to T transition mutations.[6,7] Oxidative damage to DNA by reactive oxygen species, which arise as a result of normal oxidative metabolism, is also an important source of endogenous damage. Oxidation of DNA bases occurs predominantly through radical addition to the π-bonds of the purine or pyrimidine rings.[8] The products of such reactions, e.g., 8-oxoguanine, formamidopyrimidines and ring saturated pyrimidines such as thymine glycol, are often mutagenic or cytotoxic.[9]

Eukaryotic DNA Damage Surveillance and Repair, edited by Keith W. Caldecott.

The ubiquity of DNA damaging agents and the need to ensure the integrity of the genome has necessitated the evolution of a system of DNA repair mechanisms which are able to reverse the effects of DNA damage. The majority of the simpler, nonbulky base lesions discussed above are repaired via Base Excision Repair (BER) pathways[10] that have been functionally conserved throughout evolution and are present in all forms of life from bacteria to higher mammals.[11] This review will focus primarily on mammalian BER, and in particular we will discuss orchestration of the repair process and coordination of BER pathways. However, to set the stage, we will give some background information on proteins involved in base excision repair.

Base Excision Repair Proteins

DNA Glycosylases

The basic enzymology of BER has been reasonably well elucidated and the repair pathways have been reconstituted in vitro using purified enzymes.[12-18] BER is a multistep process initiated by the activity of a class of enzymes, termed DNA glycosylases, which specifically catalyze the removal of the damaged base through cleavage of the C1'-N glycosylic bond, leaving an AP site. The first such enzyme to be identified was uracil-DNA glycosylase (UNG), discovered by T. Lindahl[19] in the mid 1970s following the correct hypothesis that life must have evolved some mechanism to deal with uracil arising through the spontaneous hydrolytic deamination of cytosine. It was subsequently discovered that UNG was only a single representative of a class of enzymes that are responsible for the removal of a wide variety of simple base damages (see Table 1). The range of DNA glycosylases, each with its own specificity, reflects their unique role in identifying a broad spectrum of DNA base lesions and initiating BER process. Known DNA glycosylases effectively cover repair of the major classes of DNA lesions caused by hydrolysis, oxidation and nonenzymatic methylation.[1] DNA glycosylases have been functionally conserved throughout evolution as a result of their importance in maintaining genome stability, and much of what is now known about the mode of action of DNA glycosylases in mammalian systems has been derived from structural and biochemical studies on homologous enzymes from *E. coli.*

High-resolution structures of a number of DNA glycosylases both native[20-25] and complexed to DNA[20,26-36] have been obtained in recent years and have provided an insight into how these enzymes overcome the significant challenge of recognizing small base modifications in the presence of a vast excess of unmodified bases. Although DNA glycosylases share little sequence homology they divide into only two structural subfamilies, the UNG family, comprising UNG, TDG, SMUG1 and MUG, and the helix-hairpin-helix family which contains such DNA glycosylases as MBD4, OGG1, MYH and NTH1.[37] All DNA glycosylases characterized thus far use nucleotide flipping to facilitate access by the active site to the anomeric center of the target nucleotide, which would otherwise be inaccessibly buried in the DNA structure. Flipping is initiated through minor groove contacts and the local structure is subsequently stabilized by insertion of one or more amino acids into the helix at the orphan base. The flipped out base is inserted into the active site pocket of the DNA glycosylase which contains aromatic residues which stack against the base, polar residues which induce specificity through hydrogen bonding contacts and hydrophobic residues that provide steric control of binding. The differing specificity of each glycosylase can therefore be understood in these terms.

UNG, for example, has extremely narrow substrate specificity, recognizing only uracil. This is due to the restricted size of the pocket which is too small for either purines or thymine to bind, with cytosine being excluded through repulsive interactions with polar groups complementary to uracil.[38] Other glycosylases have broad specificity and recognize and remove many

Table 1. Mammalian DNA glycosylases

Glycosylase	Bifunctional?	Substrates	References
UNG	No	U UNG1-mitochondrial UNG2-nuclear	201-203
SMUG1	No	U, OHMeU	204,205
TDG	No	U:G, T:G, εC	100,206
MBD4	No	U:G, T:G in CpG sites	207,208
OGG1	Yes	8oxoG:C, FapyA, FapyG	214-216
MYH	No	A:8oxoG	209,210
NTH1	Yes	Oxidised and fragmented pyrimidines (e.g., Tg, DHU), Fapy	211-213
AAG (MPG)	No	3-meA, 7-meG, εA, εG, Hx	39,217,218
NEIL1 (NEH1)	Yes	FapyA, FapyG, DHU, Tg, 8-oxoG (low)	50,56,58
NEIL2 (NEH2)	Yes	Cytosine oxidation products (5-OHU, DHU)	57

different related base damages. 3-methyladenine DNA glycosylase (AAG), for example, is able to excise modified purines as diverse as 3-methyladenine (3meA), 7-methylguanine (7meG), hypoxanthine (Hx) and ethenoadenine (εA).[39-44] In this instance, the active site pocket is lined with aromatic residues and it may be the case that the delocalized positive charge on N-alkylated purines interacts with an aromatic side chain to form a high binding energy π-donor-acceptor pair.[29]

Glycosylases can also be broadly divided into two different mechanistic subclasses: the monofunctional or "pure" glycosylases and the bifunctional glycosylases, which have an associated AP lyase activity that cleaves the phosphodiester backbone 3' to the abasic site. In the former case, the base is displaced by nucleophilic attack at the anomeric center by an activated water molecule whereas in the latter, an amino group on the enzyme acts as the nucleophile thereby forming a transient Schiff base intermediate. This Schiff base is primed to undergo a β-elimination reaction resulting in cleavage of the phosphodiester backbone, generating a 5'-phosphate and a 3'-α,β-unsaturated aldehyde. Alternatively, the Schiff base can undergo hydrolysis to leave the phosphodiester backbone intact. The degree to which the glycosylase and lyase activities of bifunctional glycosylases are coupled depends on both the particular enzyme and reaction conditions. The lyase activity of OGG1 has been estimated to be an order of magnitude less than its glycosylase activity.[45]

Mouse knockout models have been constructed for AAG,[43,46] OGG1,[47,48] UNG,[49] NTH[50] and MBD4;[51] interestingly, these mice are viable and display no overt phenotypic abnormalities. In contrast, mice deficient in genes involved in the later damage processing steps of BER, such as APE1,[52] XRCC1[53] and Pol β[54] are embryonic lethal. This could be due to cells being able to cope with the persistence of one type of damage in the genome more efficiently than the loss of repair of all AP sites, either spontaneously occurring or arising through glycosylase activity. An alternative is that glycosylases may possess a certain amount of functional redundancy, allowing other glycosylases to serve as back-ups for repair of low levels of damage. It has been shown that SMUG1 is able to act as a backup in UNG-deficient mice[55] and this may explain why these mice do not have the mutator phenotype associated with UNG-deficient bacteria and yeast.[49] Further evidence for functional redundancy among glycosylases has been provided by the identification of human homologues of the *Escherichia coli* enzyme, endonuclease VIII (NEIL1 and NEIL2)[56-59] and the isolation from mouse extracts of two additional thymine glycol-DNA glycosylase activities.[59] Between them, these enzymes can remove a wide range of oxidative lesions which were previously believed to be the preserve of OGG1 and NTH1 (see Table 1).

The redundancy in substrate specificity of certain glycosylases may reflect differing roles played within the cell cycle. UNG and MYH have each been demonstrated to associate with the replication factors RPA and PCNA and show much higher activity in proliferating tissue: these enzymes may therefore act at the replication fork,[60,61] removing damaged nucleotides misincorporated by polymerases. In contrast, other glycosylases, such as SMUG1, are expressed constitutively and most probably function in general repair of damage to genomic DNA. It was recently suggested that in human cells UNG2 (nuclear form of UNG) is the major repair enzyme for the removal of uracil arising from cytosine deamination or misincorporation with SMUG1 as a broad specificity backup.[62] Interestingly, UNG2 and SMUG1 are the only known DNA glycosylases that are stimulated by Mg^{2+} and may remove uracil from both single- and double-stranded DNA.[62]

The activity of DNA glycosylases is modulated by protein-protein interactions with other enzymes that are likely to be significant in vivo. Furthermore, it seems increasingly likely that post-translational modification of glycosylases plays an important role in their function. Phosphorylation of both MYH[63] and OGG1[64] and sumo-ylation of TDG[65] have been reported to occur in vivo.

AP Endonucleases

The recognition and removal of damaged or incorrect bases from DNA by DNA glycosylases leads to the formation of AP sites. AP sites also arise spontaneously at a substantial rate.[66,67] In addition to this, damaging chemicals such as free radicals and alkylating agents promote the release of bases, often by introducing base modifications that destabilize the N-glycosyl bond by generating a better leaving group moiety.[1,2] AP sites are a threat to cellular viability and genomic integrity as they can prevent normal DNA replication and are cytotoxic and mutagenic,[3,4,68] which is why the cell contains systems to identify and repair such sites in DNA.

A critical step in the BER pathway involves the cleavage of baseless sites in DNA by an AP endonuclease. The major AP endonuclease in mammalian cells is APE1 also known as Ref1 or HAP1, an enzyme that is homologous to bacterial exonuclease III.[69,70] APE1 is a multifunctional protein that is not only responsible for repair of AP sites, but also functions as a reduction-oxidation (redox) factor maintaining transcription factors in an active reduced state, and has been shown to be closely linked to apoptosis (reviewed in refs. 71-72). The functional importance of APE1 is underscored by the findings that mice nullizygous for the APE1 gene are embryonic lethal.[52]

APE1 is the major human AP endonuclease that comprises more than 95% of cellular AP site incision activity[73] and it cleaves phosphodiester bonds hydrolytically in a Mg^{2+}-dependent manner, leaving a 3'-hydroxyl group and a 5'-deoxyribose phosphate (dRP) group flanking the nucleotide gap. The APE1 protein is modest in size; 317 amino acids in length and approximately 35kDa. It contains two distinct domains; the N-terminal domain contains the nuclear localization sequence and is essential for redox activity, while the endonuclease activity resides in the C-terminal region.[74] APE1 cleaves abasic sites in double-stranded DNA far more efficiently than in single-stranded substrates.[75] However, the ring structure of an abasic site, the opposing base and the helical conformation prior to binding are not critical components for APE1 complex formation.[76,77] It has been postulated that since DNA is already prone to deformation by the presence of an abasic site that APE1 may bind to DNA and induce a conformational change.[78] In support of this hypothesis, it has been shown that APE1-bound DNA is distorted.[79] This bending of DNA could enable the abasic site to flip out of the DNA and move into position within the active site.[79,80]

In addition to AP endonuclease activity, APE1 has 3'-repair diesterase or phosphatase activity, albeit at approximately 100-fold lower than its AP endonuclease activity.[72,73] Although APE1 was proposed to be responsible for removing a variety of 3'-blocking groups including 3'-phosphate or 3'-phosphoglycolate that occur at single strand breaks created by oxidative agents, its poor activity suggests alternative repair mechanisms for such substrates.[9] Additionally, it has recently been shown that APE1 has a DNA exonuclease activity on mismatched deoxyribonucleotides at the 3'-termini of nicked or gapped DNA molecules. This finding suggests that APE1 may increase the fidelity of base excision repair by correcting insertion errors made by DNA polymerase β, which lacks its own proofreading activity.[81]

The complete APE1 gene is 2.6 kb and consists of four introns and five exons, the first of which is noncoding. The APE1 promoter lacks a TATA box and has multiple transcription start sites.[82-84] The upstream region of the promoter contains negative regulatory elements: one negative calcium-responsive element-A type (nCaREA) sequence and two nCaREB type sequences: B1 and B2.[85] APE1 expression is ubiquitous, however, it exhibits a complex and heterogeneous staining pattern that differs among tissue types and even differs between neighboring cells.[86,87] As expected, APE1 is found in the nucleus of cells which corresponds to its repair functions. In some cell types, however, APE1 is also found in the cytoplasm which may correspond either to its role as a DNA repair protein in the mitochondria[88] or, through its redox activity, in maintaining newly synthesized transcription factors in a reduced state during transportation to the nucleus.[71,89]

APE1 is regulated at both the transcriptional and post-translational level. Oxidative agents induce APE1, characterized by a transient increase in APE1 protein and mRNA.[90-92] This reactive oxygen species-mediated increase in protein and mRNA has been shown to involve de novo synthesis since various agents that block protein synthesis also block the induction of APE1.[92] Additionally, induction of APE1 is associated with an increase in AP endonuclease activity and an increase in resistance to H_2O_2, methyl methane sulfonate, bleomycin, γ-radiation, but not UV light.[90,92] Such resistance is also transient and is consistent with the reduction in APE1 mRNA and protein levels.[92]

Mammals also possess a second protein (APE2) with sequence homology to exonuclease III, but this protein exhibits comparatively weak AP site-specific and 3'-nuclease activities,[93] despite having a highly conserved nuclease domain with APE1. APE1 and APE2 do however contain distinguishing amino and carboxyl-terminal ends, APE2 was found to contain a mitochondrial targeting sequence and was also found to localize in the mitochondria suggesting that it may play a role in mitochondrial base excision repair.[94]

DNA Polymerases

There are currently at least 15 DNA polymerases in humans, exceeding the number found in any other organism.[95] Since resynthesis of a DNA sequence after excision of damaged DNA is catalyzed by a DNA polymerase, these enzymes clearly play a central role in DNA repair and hence in genomic stability. Table 2 details some of the major DNA polymerases, two of which, Pol β and Pol δ, have been shown to be actively involved in BER.

Table 2. Mammalian DNA polymerases

DNA Pol	Human	Activity	Polymerase family
Pol α	Pol α. Four subunits, 165kDa, 67-86kDa, 58kDa and 48kDa	DNA replication	B family of polymerases
Pol β	Pol β. 39.0kDa, a 31kDa DNA polymerase domain and an 8kDa dRPase domain	BER, both short-patch and long-patch. Neural development. Implicated in meiosis and NER	X family of polymerases
Pol δ	Pol δ. Catalytic subunit 124kDa, regulatory subunit 51.2kDa, small subunit 12.4kDa	Long patch BER, NER & MMR	Catalytic subunit belongs to the B family of polymerases
Pol ε	Pol ε. Catalytic subunit 258kDa, regulatory subunit 59.5kDa	Long patch BER, NER & MMR	Catalytic subunit belongs to the B family of polymerases
Pol γ	Pol γ. 140kDa, a heterotrimer of catalytic subunit plus homodimer of accessory subunits	Mitochondrial DNA replication and BER in mitochondrial DNA	A family of polymerases
Pol η (XPV) (RAD30A)	Pol η. 78.3kDa. Structural and functional homologue of yeast pol η	Error-free bypass of thymine dimers and other error-free and error-prone lesion bypass	UmuC/DinB superfamily, Y family
Pol ι (RAD30 B)	Pol ι. 80.3kDa	Lesion bypass	UmuC/DinB superfamily, Y family
Pol κ (DINB1)	Pol κ. 99kDa	Lesion bypass	UmuC/DinB superfamily, Y family
Pol λ	Pol λ. 39.0kDa, a 31kDa DNA polymerase domain and an 8kDa dRPase domain	Potentially functions in base excision repair or meiotic recombination, has similar properties to Pol β	X family of polymerases
Pol ζ	Pol ζ	Lesion bypass, extension from mismatched pairs	B family of polymerases

The major pathway of base excision repair, short-patch BER, uses Pol β for repair of nuclear DNA. DNA polymerase β, a member of the X family of DNA polymerases, is the lowest molecular mass DNA polymerase, being a single polypeptide of 39kDa comprising 335 amino acid residues. Its structure is highly conserved in mammals,[96] consisting of two domains connected by a protease-sensitive hinge region. The 8-kDa N-terminal domain carries dRP lyase activity (to remove the 5'-deoxyribose phosphate from an abasic site generated by AP endonuclease) and template binding functions, whereas the 31-kDa domain carries the DNA polymerase activity.[97]

The crystal structure of the rat Pol β enzyme revealed a typical polymerase fold, described as a thumb, palm and fingers, due to its resemblance to the human hand.[98,99] However, in contrast to other polymerases, Pol β lacks intrinsic accessory activities, such as 3'- or 5'-exonuclease, endonuclease, dNMP turnover and pyrophosphorolysis.[96] In vitro inhibitor studies provided the first indication of a role for Pol β in short-patch BER.[100,101] Additional evidence for the involvement of Pol β in base excision repair came from genetic studies of Pol β-deleted cells, which exhibited increased sensitivity to monofunctional DNA-alkylating agents and were found to be defective in uracil-initiated short-patch base excision repair.[16,102] The dRP lyase activity of Pol β is important in short-patch BER and it is most likely that Pol β is the major dRP lyase in mammalian cells, and other DNA polymerases with dRP lyase activity are not able to provide a reasonable backup for Pol β dRP lyase.[103,104] Further support for a role of Pol β in base excision repair comes from the fact that it is able to associate with a number of other base excision repair enzymes including DNA ligase I,[105] AP endonuclease and the XRCC1-DNA ligase III complex.[13,106]

Pol β has been present in all tissues examined and is generally present at only low levels in cells. It is constitutively expressed and is not regulated in a cell-cycle specific manner.[96] However, expression of mammalian Pol β genes is upregulated by treatment of cells with monofunctional DNA alkylating and oxidative stress-inducing agents.[107,108]

DNA polymerase δ (Pol δ), the principal DNA replicase in eukaryotes, also participates in several DNA repair pathways, including nucleotide excision repair, mismatch repair and long-patch base excision repair.[109-111] Mammalian DNA Pol δ has four subunits of 125, 66, 50 and 12 kDa.[112,113] The 125kDa catalytic subunit contains the DNA polymerase and 3' to 5' exonuclease active sites, while the role of the other subunits is unclear.[114]

For both replication and repair functions, Pol δ requires an accessory protein, the proliferating cell nuclear antigen (PCNA) to carry out highly processive DNA synthesis.[115] PCNA stimulates Pol δ by enhancing processivity, the rate-limiting step for Pol δ-catalyzed dNMP polymerization.[116] PCNA enhances processivity both by modestly increasing the rate of single nucleotide incorporation[117] and by dramatically decreasing the dissociation of Pol δ from model template-primers.[118] It is thought that PCNA performs this latter function by acting as a sliding clamp that tethers Pol δ to DNA during replication.[119]

DNA Ligases

The penultimate steps of both the short- and long-patch BER pathways result in formation of a nicked site which must be sealed by a DNA ligase for repair to be completed. Two DNA ligases have been implicated in performing this role: DNA ligase I and DNA ligase III.[120]

DNA ligase I (102 kDa) is conserved in all eukaryotes, with orthologs being found in organisms from yeast to mammals [121] and plays an essential role in DNA replication.[122] DNA ligase I associates with replication forks,[123] where it is active in joining Okazaki fragments,[124] and is up-regulated during cell proliferation.[125] In addition to its replicative role, DNA ligase I has also been implicated in several DNA repair pathways including both nucleotide excision repair[126] and base excision repair.[127] Human cell-lines containing partially inactive DNA ligase I mutants exhibit hypersensitivity to such DNA-damaging agents as ionizing radiation and alkylating agents,[128-130] suggesting a role for DNA ligase I in BER. However, a mouse model with an identical mutation was defective in DNA replication, but not in DNA repair.[131] Moreover, DNA ligase I was found to be essential for normal development, but not essential for DNA repair.[132,133] On the other hand, there is evidence to suggest the involvement of DNA ligase in the long-patch pathway of BER. In addition to the observed deficiency in long-patch BER of the human DNA ligase I mutant, 46BR.1G1,[127] in vitro studies demonstrated in particular that the interaction between PCNA and DNA ligase I plays a key role in the long-patch pathway.[127] Interaction with PCNA stimulates DNA ligase I activity through increased binding to nicked substrates.[134] Further support for DNA ligase I being involved in long-patch repair derives from the stimulation of its catalytic efficiency by replication protein A (RPA)[135] which has also been shown to specifically stimulate long-patch BER in both mammalian cell extracts[136] and a reconstituted system.[137]

So, there is no consistency in a current data on the role DNA ligase I in BER. Some data suggest DNA ligase I as the major DNA ligase for long-patch repair, other data suggesting only a minor role of this DNA ligase in BER. These two apparently contradictory sets of data clearly indicates that further investigation is required for the role of DNA ligase I in BER to be successfully unraveled.

DNA ligase III exists in the nucleus as two isoforms, denoted DNA ligase IIIα (103 kDa) and DNA ligase IIIβ (96 kDa), which arise through differential splicing of RNA transcripts.[138] DNA ligase IIIα is ubiquitously expressed, whereas DNA ligase IIIβ is found only in testes and is therefore believed to be involved in homologous recombination during meiosis.[121] The two isoforms share many features in common and differ only at their C-terminii, with the C-terminal 77 amino acids of DNA ligase IIIα being replaced with 17 different amino acids in DNA ligase IIIβ.[139] Both contain a putative zinc finger motif homologous to that of PARP-1,[140,141] and this is believed to function as a nick sensor.[106] However this motif does not seem to be essential for activity on simple nicks[140] and may instead be required for carrying out ligation at sites with complex secondary structure[142] or ligating nicks at physiological salt concentrations.[143]

DNA ligase IIIα was originally purified in complex with XRCC1.[144,145] This interaction has been characterized in vitro and is mediated through the BRCT domain located at the C-terminus of DNA ligase III.[146] XRCC1 null cell lines have 4-6 fold reduced levels and activity of DNA ligase III, indicating that the interaction may enhance the stability of the ligase[145] in addition to potentially modulating its activity. In support of this finding it has been shown that the activity of DNA ligase III can be restored in mutant XRCC1 cells (EM9) by the XRCC1 gene product.[147] Since XRCC1 null cells are characteristically deficient in short-patch BER, it has been suggested that the XRCC1-ligase III complex plays an important role in this sub-pathway,[148] although biochemical data have suggested that either DNA ligase III or DNA ligase I could be active in this pathway.[13,105]

XRCC1

Whilst XRCC1 has been strongly implicated in BER, its precise function is not yet fully understood. The *XRCC1* gene was first identified through its ability to restore a normal phenotype to the Chinese hamster ovary (CHO) cell mutant EM9, which is hypersensitive to alkylating agents and exhibits defective strand break rejoining,[149] This evidence suggested the involvement of the XRCC1 protein (69.5 kDa) in BER. XRCC1 has since been shown to interact with a number of BER proteins including DNA ligase IIIα, Pol β, polynucleotide kinase and PARP-1.[13,106,144,146,150,151] The N-terminus of XRCC1 interacts with the palm-thumb domain of Pol β allowing it to form a tertiary complex with Pol β and nicked DNA, in which both proteins almost completely surround the nicked site. XRCC1 binds to the concave side of the 90° bent nicked DNA, whilst Pol β binds to the DNA's convex side, sequestering the damage from the cellular milieu. The N-terminal domain of XRCC1 has additionally been demonstrated to have nick-binding activity.[152]

XRCC1 has two BRCT domains, the first (BRCT I) of which is involved in its interaction with PARP-1 while the second (C-terminal) domain (BRCT II) is responsible for binding DNA ligase III.[153,154] Mutations in the BRCT domains of XRCC1 which prevent them from interacting with either DNA ligase IIIα or PARP-1 seriously compromise single strand break repair in vivo. The BRCT II domain seems to be required only for single strand break repair during the G_0/G_1 phase of the cell cycle [155,156] while the BRCT I domain is required for single strand break repair in both G_1 and S/G_2.[157]

XRCC1 has also been shown to interact with human polynucleotide kinase, forming a multiprotein complex with Pol β and DNA ligase III.[150] XRCC1 stimulates both the DNA kinase and DNA phosphatase activities of human polynucleotide kinase at damaged DNA termini and accelerates the repair process.[150] A model has therefore been proposed whereby activated PARP-1 interacts with XRCC1 at a nick-site, XRCC1 then binds to the concave side of the DNA acting as a scaffold for other proteins to access the nick.[149] Recent research suggests that XRCC1 may become involved even earlier in the BER pathway than had previously been thought. XRCC1 is able to stimulate turnover of APE1 suggesting that XRCC1 becomes involved at the incision step of BER.[158]

The *XRCC1* gene is expressed at reasonably low levels in most tissues, however the mRNA level is higher in testes. This expression pattern suggests that XRCC1 plays some role in meiotic recombination events, however the testes-specific DNA ligase IIIβ does not interact with XRCC1.[159] The observed embryonic lethality of deletion of the *XRCC1*[53] gene may therefore reflect its importance in maintaining genomic stability or may instead be due to some other as yet undefined role of XRCC1.

Flap Endonuclease 1

Flap endonuclease 1 (FEN1) is a 42kDa Mg^{2+}-dependent exo- and endonuclease that specifically recognises a 5'-single-stranded flap, tracks down the flap to the point where the two strands of duplex DNA adjoin the single-stranded flap and cleaves the phosphodiester bond next to it.[160] FEN1 is involved in lagging strand DNA synthesis, in Okazaki fragment joining, and is also involved in DNA repair. FEN1 protein has a nuclear localization signal and in mammalian cells is promptly relocated to the nucleus after DNA damage.[161,162] Deletion of the FEN1 homologue in the yeast *S. cerevisiae* and *S. pombe* resulted in marked sensitivity to the alkylating agent methyl-methane sulfonate and chromosome instability.[163,164] Surprisingly, although hypersensitive to DNA damaging agents, vertebrate cells lacking FEN1 are viable, but exhibit a slow growth phenotype.[165]

Major evidence for the role of FEN1 in long-patch BER arose from in vitro experiments. In a repair system reconstituted with purified enzymes Klungland and Lindahl demonstrated that FEN1 was essential for repair of reduced AP sites.[14] It was also shown that immunodepletion of FEN1 from human whole cell extract significantly decreases the efficiency of long-patch BER, suggesting that FEN1 is the major flap activity involved in this repair subpathway.[166] FEN1 is not able to remove the 5'-sugar phosphate generated by the cleavage of an AP site by APE1, thus the number of nucleotides removed by FEN1 during long-patch repair must be at least two.[15,167] PCNA interacts with FEN1[168,169] and stimulates its flap endonuclease activity through this interaction: a PCNA mutant that could not bind to FEN1 was unable to stimulate excision.[170] FEN1 also interacts with APE1 and Werner protein, but the physiological significance of these interactions is unclear.[171,172]

Poly(ADP-Ribose)Polymerase

The possibility that the enzyme poly(ADP-ribose)polymerase (PARP-1) is an active participant in BER in vivo has been extensively investigated in recent years but remains controversial. PARP-1 binds to single strand breaks and subsequently catalyses the modification of nuclear proteins with polymers of ADP-ribose which are synthesized from NAD^+. PARP-1 null mice are hypersensitive to alkylating agents and ionising radiation,[173,174] and immortalised cells from these animals manifest characteristic features of genomic instability, including delayed rejoining of strand breaks, mixed ploidy and elevated frequencies of sister chromatid exchange and micronuclei formation upon treatment with DNA-damaging agents.[173-175] Strand break repair exhibits a dependence on NAD^+ which has been attributed to the activity of PARP-1.[176,177] One of the major targets of poly(ADP-ribosyl)ation is PARP-1 itself. PARP-1 binds to strand breaks as a homodimer and after automodification, resulting in accumulation of negative charge on the enzyme, it dissociates from the DNA. Recycling of PARP-1 is provided by PAR-glycohydrolase which catalyzes degradation of ADP-ribosyl polymers attached to PARP.[178] A model was proposed whereby PARP-1 plays a passive role in BER, temporarily binding to nicked DNA repair intermediates to protect them from degradation by the cellular milieu.[177] Recent studies of repair by cell extracts from PARP-1-deficient mouse fibroblasts would seem to support this theory. PARP-1 deficient cell extracts showed similar ability to repair X-irradiated or alkylating agent-treated plasmid DNA compared to wild-type cell extracts.[179] Similarly, studies carried out using DNA substrates containing a site-specifically located normal AP site or reduced AP site (long-patch repair substrate) showed that repair in PARP-1 deficient extracts was more efficient than in extracts from the wild-type.[219] In both studies titration of PARP-1 into the PARP-1-deficient extracts inhibited rather than stimulated repair.[179,219] However, an alternative theory, that PARP-1 plays an active role in DNA damage processing, has also been advanced. Experiments with purified recombinant enzymes demonstrated stimulation of Pol β by PARP-1, although only in the absence of NAD^+.[180] It has also been reported that extracts from PARP-1-deficient mouse fibroblasts show decreased long-patch synthesis in the repair of a DNA substrate containing an abasic site and it was therefore concluded that PARP-1 is required for long-patch BER.[181] Clearly, further work is required to understand the role of PARP in BER.

Coordination of Base Excision Repair

Short-Patch Repair Pathway

As we already mentioned, glycosylases can be divided into two different mechanistic subgroups depending of whether they have an associated AP lyase activity.[75] Correspondingly, there are two possible scenarios for short-patch repair initiated by different glycosylases. In the case of monofunctional DNA glycosylases, of which uracil-DNA glycosylase (UNG), is a classic example, the base is removed through nucleophilic attack of the N-glycosylic bond by an activated water molecule. Repair then proceeds via a highly coordinated series of concerted reactions according to the 'passing-the baton' model of BER,[11,182] (Fig. 1, left branch). It has been demonstrated that UNG remains associated with the AP site [35] following removal of uracil, protecting the AP site from degradation. The glycosylase is subsequently displaced by the next enzyme in the pathway, APE1, which achieves this displacement through having a greater affinity for AP sites by making extensive contacts with the DNA and inducing significant kinking of the helical axis.[79] This is supported by the observed stimulation of UNG turnover in the presence of APE1.[183,184] APE1 hydrolyses the phosphodiester backbone 5' to the AP site, leaving a 3'-OH and a 5'-deoxyribosephosphate residue (dRP).[69,185] Pol β recognizes the nicked AP site and adds a single nucleotide by displacing the 5'-terminal sugar phosphate. The 5'-dRP residue must be removed for single-nucleotide patch repair to proceed to completion. In mammals Pol β performs this task through a dRP lyase activity additional to its polymerase function.[97] The two activities are likely to be tightly coupled, since co-complementation of cell extracts from Pol β null mouse embryonic fibroblasts with both dRP lyase and polymerase deficient mutants of Pol β are not able to restore single-nucleotide patch repair to these extracts.[104]

In practice, Pol β has been demonstrated to have an affinity for the complex between APE1 and an uncleaved AP site and is recruited to the damage site as a ternary complex.[186] Following backbone cleavage by APE1 Pol β retains affinity for the DNA and is then able to both remove the dRP and introduce a single nucleotide in a concerted fashion. Since Pol β also

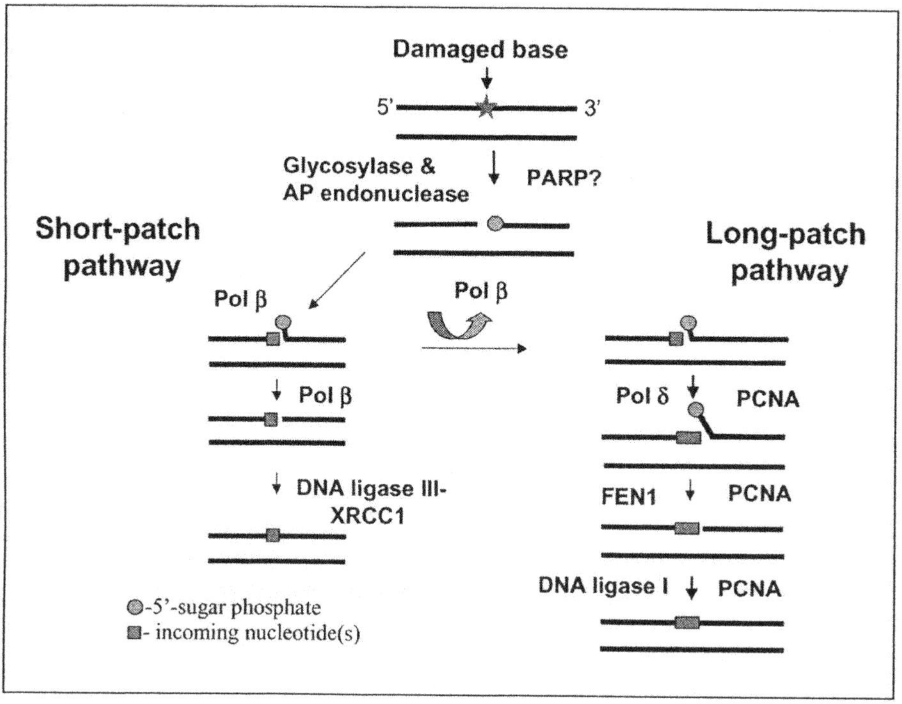

Figure 1. Model for co-ordination of BER by protein-protein interactions

has an affinity for both DNA ligase I[105,187] and DNA ligase III (mediated by XRCC1)[144,146] it is therefore possible for the final step of repair, i.e., ligation of the nick, to be coupled to the rest of the repair process. Indeed, when short-patch BER was reconstituted with purified mammalian proteins both DNA ligase III and DNA ligase I were able to accomplish repair.[13,16] The second class of glycosylases, the bifunctional glycosylases, include the oxidative damage processing enzymes hOGG1, which removes oxidized purines, and hNTH1, which removes oxidized pyrimidines. Bifunctional glycosylases remove damaged base through nucleophilic attack by an active site amine group, forming a Schiff base with C1' of the deoxyribose during the course of the excision reaction. This covalent complex can either hydrolyze, leaving an intact AP site, or can undergo β-elimination, cleaving the phosphodiester backbone 3' to the AP site, leaving a 3'-α,β-unsaturated aldehyde residue and a 5'-phosphate.[188] The blocking unsaturated aldehyde residue can then be excised by APE1,[75] although the 3'-phosphodiesterase activity of APE1 is significantly (100-fold) weaker than its AP endonuclease activity.[72] The product of sequential AP lyase and 3'-phosphodiesterase processing of an AP site is a one-nucleotide gap with flanking 3'-OH and 5'-phosphate terminii. Repair could then be completed simply by insertion of a single nucleotide followed by ligation and indeed, short-patch BER predominates in the repair of the oxidative lesions, 8-oxoguanine and thymine glycol.[189-191]

As has been demonstrable for the monofunctional glycosylases UNG and TDG[35,192] the glycosylase activity of OGG1 is stimulated by the presence of APE1.[183,184] This stimulation seems to be selective for the glycosylase activity with the associated AP lyase activity remaining unstimulated. This may have profound consequences for the relevance of the AP lyase activity of glycosylases in vivo. Recent work carried out in our laboratory has found that repair by whole cell extracts of a single 8-oxoguanine residue site-specifically incorporated into closed circular substrate DNA proceeds via a Pol β-dependent short-patch mechanism. In the absence of Pol β, i.e., in Pol β null mouse cell extracts, repair is only able to proceed through the long-patch pathway. Complementation of Pol β null extracts with the dRP lyase-deficient Pol β K72A mutant did not restore single-nucleotide patch repair. These data suggested that a dRP residue remained at the 5' terminus following processing of 8-oxoguanine by OGG1 and APE1 with the implication that repair proceeded via the same pathway as that for repair initiated by monofunctional glycosylases.[16]

It is tempting to speculate that the majority of short-patch BER proceeds through a single general mechanism regardless of the AP lyase activity of the initiating glycosylase, however this subject requires further clarification.

Long-Patch BER Pathway

As discussed above, the dRP lyase activity of Pol β must remove the dRP residue from the 5'-terminus of the nicked BER intermediate before ligation and therefore completion of the short-patch repair pathway can occur. However, the dRP lyase activity of Pol β proceeds via a β-elimination mechanism and is therefore exquisitely sensitive to chemical modification of the dRP. Both oxidized[193] and reduced AP sites[194] are refractory to β-elimination and repair of these sites must occur by an alternative mechanism. This repair is accomplished through the long- patch BER pathway[14,111] in which between 2 and 8 nucleotides are incorporated into the repair gap. The dRP is displaced as part of an oligonucleotide 'flap' which is subsequently removed by FEN1.[14,195] DNA ligase is then free to complete the repair process by rejoining the free ends of the DNA.

The long-patch pathway has been reconstituted in vitro using purified proteins by a number of laboratories and it has been demonstrated that under such conditions both Pol β and Pol δ are able to support repair.[14,15,17,194] Long-patch repair was found to be proliferating cell nuclear antigen (PCNA) dependent and a model was proposed whereby Pol δ, the activity of which is PCNA dependent, was responsible for the strand resynthesis step of long-patch BER.[111,196] This model, however, prompts the question: how does the cellular repair machinery direct repair through one or other pathway before the dRP intermediate is reached? As discussed above, the individual steps of BER proceed via a highly orchestrated mechanism in which each enzyme interacts directly with the next, passing the substrate along until repair is completed. If Pol δ were responsible for initiating strand resynthesis in long-patch BER, it would require that the modified AP site be identified as such before recruitment of the polymerase to the repair complex.

Interestingly, it was reported by several groups that addition of Pol β-neutralizing antibodies suppressed long-patch repair of reduced AP sites by human whole cell extracts.[14,197] The finding that either Pol β or Pol δ may be involved in long-patch BER, raised the questions of why two DNA polymerases should be involved and what determines which one is engaged? Recently we proposed a model suggesting a solution to this dilemma. We have demonstrated that both short- and long-patch BER proceed via a unified mechanism up to insertion of the first nucleotide supported by Pol β. Unmodified AP sites are subsequently removed by the dRP lyase activity of Pol β allowing completion of repair by the short-patch mechanism whereas β-elimination resistant AP sites would proceed through the long-patch BER pathway (Fig. 1, right branch).

The PCNA dependence of long-patch BER may imply that Pol δ contributes to strand resynthesis following insertion of the first nucleotide. Indeed, we have recently found that the polymerases switch after addition of the first nucleotide by Pol β, and Pol δ continues strand-displacement repair synthesis.[219,220] It is also worth mentioning that PCNA plays multiple roles in long-patch BER. In addition to being a processivity factor for Pol δ, PCNA also interacts with other components of the BER pathway. FEN1 and DNA ligase I each have

PCNA-binding sites[198] and PCNA has been demonstrated to stimulate both flap removal by FEN1[14,170] and ligation by DNA ligase I [134,199]

There is increasing evidence that long-patch BER also proceeds through a highly orchestrated mechanism with each step coordinated with the next through protein-protein interactions. PCNA may play a key role in such a coordination through its interaction with Pol δ,[200] FEN1[168,169] and DNA ligase I.[124] Coordination of repair in this manner is likely to have evolved to protect the genome from the potentially deleterious effects of exposing repair intermediates to the cellular milieu.

The damaged base is recognized and removed by a DNA glycosylase that then remains bound to the AP site until it is displaced by AP endonuclease which generates a strand break 5' to the abasic site. AP endonuclease interacts with Pol β and attracts it to the nicked AP site. Pol β adds one nucleotide to the 3'-end of the nick and from this point Pol β plays a central role in determining the sub-pathway choice. Removal of the 5'-sugar phosphate by dRP lyase activity of the Pol β directs BER to short-patch repair pathway that results in a single-nucleotide removal and replacement. Alternatively, Pol δ adds a few more nucleotides to the 3'-end and generates a flap containing the 5'-sugar phosphate. This switches repair to the long-patch BER sub-pathway where the flap is removed by FEN1. In the short-patch pathway DNA ligase III/ XRCC1 complex interacts with Pol β and completes DNA repair by sealing the strand break. In the long-patch pathways PCNA serves as a processivity factor for Pol δ. It also interacts with FEN1 and DNA ligase I and may stimulate both flap excision and ligation.

References

1. Lindahl T. Instability and decay of the primary structure of DNA. Nature 1993; 362:709-715.
2. Lindahl T, Andersson A. Rate of chain breakage of apurinic sites in double-stranded DNA. Biochemistry 1972; 11:3618-3623.
3. Guillet M, Boiteux S. Endogenous DNA abasic sites cause cell death in the absence of APN1, APN2 and Rad1/Rad10 in Saccharomyces cerevisiae. EMBO J 2002; 21:2833-2841.
4. Loeb LA, Preston BD. Mutagenesis by apurinic/apyrimidinic sites. Ann Rev Genet 1986; 20:201-230.
5. Lindahl T, Nyberg B. Heat-induced deamination of cytosine residues in DNA. Biochemistry 1974; 13:3405-3410.
6. Duncan BK, Miller J. Mutagenic deamination of cytosine residues in DNA. Nature 1980; 287:560-561.
7. Duncan BK, Weiss B. Specific mutator effects of ung (uracil-DNA glycosylase) mutations in Escherichia coli. J Bacteriol 1982; 151:750-755.
8. Burrows CJ, Muller JG. Oxidative nucleobase modifications leading to strand scission. Chemical Reviews 1998; 98(3):1109-1151.
9. Wallace SS. Enzymatic processing of radiation-induced free radical damage in DNA. Radiat Res 1998; 150(5 Suppl):S60-79.
10. Friedberg EC, Walker GC, Siede W. DNA Repair and Mutagenesis. Washington, DC: American Society of Microbiology Press, 1995.
11. Lindahl T, Wood RD. Quality control by DNA repair. Science 1999; 286:1897-1905.
12. Dianov G, Lindahl T. Reconstitution of the DNA base excision-repair pathway. Curr Biol 1994; 4(12):1069-1076.
13. Kubota Y, Nash RA, Klungland A et al. Reconstitution of DNA base excision-repair with purified human proteins: Interaction between DNA polymerase β and the XRCC1 protein. EMBO J 1996; 15:6662-6670.
14. Klungland A, Lindahl T. Second pathway for completion of human DNA base excision-repair: Reconstitution with purified proteins and requirement for DNase IV (FEN1). EMBO J 1997; 16(11):3341-3348.
15. Matsumoto Y, Kim K, Hurwitz J et al. Reconstitution of proliferating cell nuclear antigen-dependent repair of apurinic/apyrimidinic sites with purified human proteins. J Biol Chem 1999; 274(47):33703-33708.

16. Allinson SL, Dianova II, Dianov GL. DNA polymerase β is the major dRP lyase involved in repair of oxidative base lesions in DNA by mammalian cell extracts. EMBO J 2001; 20(23):6919-6926.
17. Pascucci B, Stucki M, Jonsson ZO et al. Long patch base excision repair with purified proteins. J Biol Chem 1999; 274:33696-33702.
18. Pascucci B, Maga G, Hubscher U et al. Reconstitution of the base excision repair pathway for 7,8-dihydro-8-oxoguanine with purified human proteins. Nucleic Acids Res 2002; 30(10):2124-2130.
19. Lindahl T. An N-glycosidase from Escherichia coli that releases free uracil from DNA containing deaminated cytosine residues. Proc Natl Acad Sci USA 1974; 71:3649-3653.
20. Bjoras M, Seeberg E, Luna L et al. Reciprocal "flipping" underlies substrate recognition and catalytic activation by the human 8-oxo-guanine DNA glycosylase. J Mol Biol 2002; 317(2):171-177.
21. Drohat AC, Kwon K, Krosky DJ et al. 3-methyladenine DNA glycosylase is an unexpected helix-hairpin-helix superfamily member. Nature Struct Biol 2002; 9:659-664.
22. Barrett TE, Savva R, Panayotou G et al. Crystal structure of a G:T/U mismatch-specific DNA glycosylase: Mismatch recognition by complementary-strand interactions. Cell 1998; 92(1):117-129.
23. Savva R, McAuley-Hecht K, Brown T et al. The Structural Basis of Specific Base-Excision Repair by Uracil-DNA Glycosylase. Nature 1995; 373(6514):487-493.
24. Thayer MM, Ahern H, Xing DX et al. Novel DNA-Binding motifs in the DNA-Repair enzyme endonuclease III crystal structure. EMBO J 1995; 14(16):4108-4120.
25. Mol CD, Arvai AS, Slupphaug G et al. Crystal-Structure and mutational analysis of human Uracil-DNA Glycosylase - Structural basis for specificity and catalysis. Cell 1995; 80(6):869-878.
26. Lau AY, Scharer OD, Samson L et al. Crystal structure of a human alkylbase-DNA repair enzyme complexed to DNA: Mechanisms for nucleotide flipping and base excision. Cell 1998; 95(2):249-258.
27. Hollis T, Ichikawa Y, Ellenberger T. DNA bending and a flip-out mechanism for base excision by the helix-hairpin-helix DNA glycosylase, Escherichia coli AlkA. EMBO J 2000; 19(4):758-766.
28. Parikh SS, Walcher G, Jones GD et al. Uracil-DNA glycosylase-DNA substrate and product structures: Conformational strain promotes catalytic efficiency by coupled stereoelectronic effects. Proc Natl Acad Sci USA 2000; 97(10):5083-5088.
29. Lau AY, Wyattt MD, Glassner BJ et al. Molecular basis for discriminating between normal and damaged bases by the human alkyladenine glycosylase, AAG. Proc Natl Acad Sci USA 2000; 97(25):13573-13578.
30. Gilboa R, Zharkov DO, Golan G et al. Structure of formamidopyrimidine-DNA glycosylase covalently complexed to DNA. J Biol Chem 2002; 277(22):19811-19816.
31. Labahn J, Scharer OD, Long A et al. Structural basis for the excision repair of alkylation-damaged DNA. Cell 1996; 86(2):321-329.
32. Bruner SD, Norman DPG, Verdine GL. Structural basis for recognition and repair of the endogenous mutagen 8-oxoguanine in DNA. Nature 2000; 403(6772):859-866.
33. Norman DPG, Bruner SD, Verdine GL. Coupling of substrate recognition and catalysis by a human base-excision DNA repair protein. J Am Chem Soc 2001; 123(2):359-360.
34. Fromme JC, Verdine GL. Structural insights into lesion recognition and repair by the bacterial 8-oxoguanine DNA glycosylase MutM. Nat Struct Biol 2002; 9(7):544-552.
35. Parikh SS, Mol CD, Slupphaug G et al. Base excision repair initiation revealed by crystal structures and binding kinetics of human uracil-DNA glycosylase with DNA. EMBO J 1998; 17(17):5214-5226.
36. Slupphaug G, Mol CD, Kavli B et al. A nucleotide-flipping mechanism from the structure of human uracil-DNA glycosylase bound to DNA. Nature 1996; 384(6604):87-92.
37. Nash HM, Bruner SD, Scharer OD et al. Cloning of a yeast 8-oxoguanine DNA glycosylase reveals the existence of a base-excision DNA-repair protein superfamily. Curr Biol 1996; 6(8):968-980.
38. Savva R, McAuley HK, Brown T et al. The structural basis of specific base-excision repair by uracil-DNA glycosylase. Nature 1995; 373(6514):487-493.
39. O'Connor TR, Laval J. Human cDNA expressing a functional DNA glycosylase excising 3-methyladenine and 7-methylguanine. Biochem Biophys Res Commun 1991; 176:1170-1177.
40. Dosanjh MK, Roy R, Mitra S et al. 1,N-6-Ethenoadenine is preferred over 3-methyladenine as substrate by a cloned human N-methylpurine DNA glycosylase (3- methyladenine DNA glycosylase). Biochemistry 1994; 33(7):1624-1628.

41. Saparbaev M, Laval J. Excision of hypoxanthine from DNA containing dIMP residues by the Escherichia coli, yeast, rat, and human alkylpurine DNA glycosylases. Proc Natl Acad Sci USA 1994; 91:5873-5877.
42. Saparbaev M, Kleibl K, Laval J. E. coli, S.cerevisiae, rat and human 3- methyladenine DNA glycosylases repair 1,N-6-ethenoadenine when present in DNA. Nucleic Acids Res 1995; 23(18):3750-3755.
43. Hang B, Singer B, Margison GP et al. Targeted deletion of alkylpurine-DNA-N-glycosylase in mice eliminates repair of 1,N_6-ethenoadenine and hypoxanthine but not of 3,N_4-ethenocytosine or 8-oxoguanine. Proc Natl Acad Sci USA 1997; 94(24):12869-12874.
44. Asaeda A, Ide H, Asagoshi K et al. Substrate specificity of human methylpurine DNA N-glycosylase. Biochemistry 2000; 39(8):1959-1965.
45. Zharkov DO, Rosenquist TA, Gerchman SE et al. Substrate specificity and reaction mechanism of murine 8- oxoguanine-DNA glycosylase. J Biol Chem 2000; 275(37):28607-28617.
46. Engelward BP, Weeda G, Wyatt MD et al. Base excision repair deficient mice lacking the Aag alkyladenine DNA glycosylase. Proc Natl Acad Sci USA 1997; 94(24):13087-13092.
47. Klungland A, Rosewell I, Hollenbach S et al. Accumulation of premutagenic DNA lesions in mice defective in removal of oxidative base damage. Proc Nat Acad Sci USA 1999; 96(23):13300-13305.
48. Minowa O, Arai T, Hirano M et al. Mmh/Oggl gene inactivation results in accumulation of 8-hydroxyguanine in mice. Proc Nat Acad Sci USA 2000; 97(8):4156-4161.
49. Nilsen H, Rosewell I, Robins P et al. Uracil-DNA glycosylase (UNG)-deficient mice reveal a primary role of the enzyme during DNA replication. Molecular Cell 2000; 5(6):1059-1065.
50. Takao M, Kanno S-i, Kobayashi K et al. A back-up glycosylase in Nth1-knockout mice is a functional Nei (endonuclease VIII) homologue. J Biol Chem 2002.in press.
51. Millar CB, Guy J, Sansom OJ et al. Enhanced CpG mutability and tumorigenesis in MBD4-deficient mice. Science 2002; 297(5580):403-405.
52. Xanthoudakis S, Smeyne RJ, Wallace JD et al. The redox/DNA repair protein, Ref-1, is essential for early embryonic development in mice. Proc Natl Acad USA 1996; 93(17):8919-8923.
53. Tebbs RS, Flannery ML, Meneses JJ et al. Requirement for the Xrccl DNA base excision repair gene during early mouse development. Dev Biol 1999; 208(2):513-529.
54. Gu H, Marth JD, Orban PC et al. Deletion of a DNA-Polymerase-β gene segment in T-Cells using Cell-Type-Specific gene targeting. Science 1994; 265(5168):103-106.
55. Nilsen H, Haushalter KA, Robins P et al. Excision of deaminated cytosine from the vertebrate genome: Role of the SMUG1 uracil-DNA glycosylase. EMBO J 2001; 20(15):4278-4286.
56. Hazra TK, Izumi T, Boldogh I et al. Identification and characterization of a human DNA glycosylase for repair of modified bases in oxidatively damaged DNA. Proc Natl Acad Sci USA 2002; 99(6):3523-3528.
57. Hazra TK, Kow YW, Hatahet Z et al. Identification and characterization of a novel human DNA glycosylase for repair of cytosine-derived lesions. J Biol Chem 2002; 277(34):30417-30420.
58. Bandaru V, Sunkara S, Wallace SS et al. A novel human DNA glycosylase that removes oxidative DNA damage and is homologous to Escherichia coli endonuclease VIII. DNA Repair 2002; 1:517-529.
59. Takao M, S-i. K, Shiromoto. et al. Novel nuclear and mitochondrial glycosylases revealed by disruption of the mouse Nthl gene encoding an endonuclease III homolog for repair of thymine glycols. EMBO J 2002; 21:3486-3493.
60. Otterlei M, Warbrick E, Nagelhus TA et al. Post-replicative base excision repair in replication foci. EMBO J 1999; 18(13):3834-8344.
61. Parker A, Gu YS, Mahoney W et al. Human homolog of the MutY repair protein (hMYH) physically interacts with proteins involved in long patch DNA base excision repair. J Biol Chem 2001; 276(8):5547-5555.
62. Kavli B, Sundheim O, Akbari M et al. hUNG2 is the major repair enzyme for removal of uracil from U: A matches, U: G mismatches and U in single stranded DNA, with hSMUG1 as a broad specificity backup. J Biol Chem 2002; 2:39926-39936.
63. Gu Y, Desai T, Gutierrez PL et al. Alteration of DNA base excision repair enzymes hMYH and hOGG1 in hydrogen peroxide resistant transformed human breast cells. Med Sci Monit 2001; 7(5):861-868.

64. Dantzer F, Luna L, Bjoras M et al. Human OGG1 undergoes serine phosphorylation and associates with the nuclear matrix and mitotic chromatin in vivo. Nucl Acids Res 2002; 30(11):2349-2357.
65. Hardeland U, Steinacher R, Jiricny J et al. Modification of the human thymine-DNA glycosylase by ubiquitin-like proteins facilitates enzymatic turnover. EMBO J 2002; 21(6):1456-1464.
66. Lindahl T, Nyberg B. Rate of depurination of native deoxyribonucleic acid. Biochemistry 1972; 11:3610-3618.
67. Lindahl T, Karlström O. Heat-induced depyrimidination of DNA. Biochemistry 1973; 12:5151-5154.
68. Gentil A, Renault G, Madzak C et al. Mutagenic properties of a unique abasic site in mammalian cells. Biochem Biophys Res Comm 1990; 173:704-710.
69. Demple B, Herman T, Chem DS. Cloning and expression of APE, the cDNA encoding the major human apurinic endonuclease: Definition of a family of DNA repair enzymes. Proc Natl Acad Sci USA 1991; 88:11450-11454.
70. Robson CN, Hickson ID. Isolation of cDNA clones encoding a human apurinic/apyrimidinic endonuclease that corrects DNA repair and mutagenesis defects in E. coli xth (exonuclease III) mutants. Nucleic Acids Res 1991; 19:5519-5523.
71. Evans AR, Limp-Foster M, Kelley MR. Going APE over ref-1. Mutat Res 2000; 461(2):83-108.
72. Wilson DM, Barsky D. The major human abasic endonuclease: Formation, consequences and repair of abasic lesions in DNA. Mutat Res 2001; 485:283-307.
73. Chen DS, Herman VG, Demple B. Two distinct human diesterases that hydrolyze 3'-blocking deoxyribose fragments from oxidised DNA. Nucleic Acids Res 1991; 19:5907-5914.
74. Xanthoudakis S, Miao GG, Curran T. The redox and DNA-repair activities of Ref-1 are encoded by nonoverlapping domains. Proc Natl Acad Sci USA 1994; 91(1):23-27.
75. Demple B, Harrison L. Repair of oxidative damage to DNA: Enzymology and biology. Annu Rev Biochem 1994; 63:915-948.
76. Erzberger JP, Barsky D, Scharer OD et al. Elements in abasic site recognition by the major human and Escherichia coli apurinic/apyrimidinic endonucleases. Nucleic Acids Res 1998; 26(11):2771-2778.
77. Wilson 3rd DM, Takeshita M, Grollman AP et al. Incision activity of human apurinic endonuclease (Ape) at abasic site analogs in DNA. J Biol Chem 1995; 270(27):16002-16007.
78. Chaudhry MA, Weinfeld M. The action of Escherichia coli endonuclease III on multiply damaged sites in DNA. J Mol Biol 1995; 249:914-922.
79. Mol CD, Izumi T, Mitra S et al. DNA-bound structures and mutants reveal abasic DNA binding by APE1 and DNA repair coordination. Nature 2000; 403:451-454.
80. Parikh SS, Mol CD, Hosfield DJ et al. Envisioning the molecular choreography of DNA base excision repair. Curr Opin Struct Biol 1999; 9(1):37-47.
81. Chou KM, Cheng YC. An exonucleolytic activity of human apurinic/apyrimidinic endonuclease on 3' mispaired DNA. Nature 2002; 415(6872):655-659.
82. Harrison L, Ascione AG, Wilson 3rd DM et al. Characterization of the promoter region of the human apurinic endonuclease gene (APE). J Biol Chem 1995; 270(10):5556-5564.
83. Robson CN, Hochhauser D, Craig R et al. Structure of the human DNA repair gene HAP1 and its localization to chromosome 14q 11.2-12. Nucleic Acids Res 1992; 20:4417-4421.
84. Zhao B, Grandy DK, Hagerup JM et al. The human gene for apurinic/apyrimidinic endonuclease (HAP1): Sequence and localization to chromosome 14 band q12. Nucleic Acids Res 1992; 20(15):4097-4098.
85. Izumi T, Henner WD, Mitra S. Negative regulation of the major human AP-endonuclease, a multifunctional protein. Biochemistry 1996; 35(47):14679-14683.
86. Wilson TM, Rivkees SA, Deutsch WA et al. Differential expression of the apurinic / apyrimidinic endonuclease (APE/ref-1) multifunctional DNA base excision repair gene during fetal development and in adult rat brain and testis. Mutat Res 1996; 362(3):237-248.
87. Rivkees SA, Kelley MR. Expression of a Multifunctional DNA-Repair Enzyme Gene, Apurinic/Apyrimidinic Endonuclease (Ape-Ref-1) in the Suprachiasmatic, Supraoptic and Paraventricular Nuclei. Brain Res 1994; 666(1):137-142.
88. Sawyer DE, Van Houten B. Repair of DNA damage in mitochondria. Mutat Res-DNA Repair 1999; 434(3):161-176.

89. Kakolyris S, Kaklamanis L, Giatromanolaki et al. Expression and subcellular localisation of human AP endonuclease 1 (HAP1/Ref-1) protein: A basis for its role in human disease. Histopathology 1998; 33:561-569.
90. Grosch S, Fritz G, Kaina B. Apurinic endonuclease (Ref-1) is induced in mammalian cells by oxidative stress and involved in clastogenic adaptation. Cancer Res 1998; 58(19):4410-4416.
91. Grosch S, Kaina B. Transcriptional activation of apurinic/apyrimidinic endonuclease (Ape, Ref-1) by oxidative stress requires CREB. Biochem Biophys Res Commun 1999; 261(3):859-863.
92. Ramana CV, Boldogh I, Izumi T et al. Activation of apurinic/apyrimidinic endonuclease in human cells by reactive oxygen species and its correlation with their adaptive response to genotoxicity of free radicals. Proc Natl Acad Sci USA 1998; 95(9):5061-5066.
93. Hadi MZ, Ginalski K, Nguyen LH et al. Determinants in nuclease specificity of Ape1 and Ape2, human homologues of Escherichia coli exonuclease III. J Mol Biol 2002; 316(3):853-866.
94. Tsuchimoto D, Sakai Y, Sakumi K et al. Human APE2 protein is mostly localized in the nuclei and to some extent in the mitochondria, while nuclear APE2 is partly associated with proliferating cell nuclear antigen. Nucleic Acids Res 2001; 29(11):2349-23460.
95. Wood RD, Mitchell M, Sgouros J et al. Human DNA repair genes. Science 2001; 291(5507):1284-1289.
96. Wilson SH. Mammalian base excision repair and DNA polymerase beta. Mutat Res 1998; 407(3):203-215.
97. Matsumoto Y, Kim K. Excision of deoxyribose phosphate residues by DNA polymerase β during DNA repair. Science 1995; 269(5224):699-702.
98. Pelletier H, Sawaya MR, Kumar A et al. Structures of Ternary Complexes of Rat DNA Polymerase β, a DNA Template-Primer, and ddctp. Science 1994; 264(5167):1891-1903.
99. Pelletier H, Sawaya MR, Wolfle W et al. Crystal structures of human DNA polymerase beta complexed with DNA: Implications for catalytic mechanism, processivity, and fidelity. Biochemistry 1996; 35(39):12742-12761.
100. Wiebauer K, Jiricny J. Mismatch-specific thymine DNA glycosylase and DNA polymerase β mediate the correction of G-T mispairs in nuclear extracts from human cells. Proc Natl Acad Sci USA 1990; 87:5842-5845.
101. Dianov G, Price A, Lindahl T. Generation of single-nucleotide repair patches following excision of uracil residues from DNA. Mol Cell Biol 1992; 12:1605-1612.
102. Sobol RW, Horton JK, Kuhn R et al. Requirement of mammalian DNA polymerase-β in base-excision repair. Nature 1996; 379:183-186.
103. Sobol RW, Prasad R, Evenski A et al. The lyase activity of the DNA repair protein β-polymerase protects from DNA-damage-induced cytotoxicity. Nature 2000; 405:807-809.
104. Podlutsky AJ, Dianova II, Wilson SH et al. DNA synthesis and dRPase activities of polymerase beta are both essential for single-nucleotide patch base excision repair in mammalian cell extracts. Biochemistry 2001; 40(3):809-813.
105. Prasad R, Singhal RK, Srivastava DK et al. Specific interaction of DNA polymerase beta and DNA ligase I in a multiprotein base excision repair complex from bovine testis. J Biol Chem 1996; 271(27):16000-16007.
106. Caldecott KW, Aoufouchi S, Johnson P et al. XRCC1 polypeptide interacts with DNA polymerase beta and possibly poly (ADP-ribose) polymerase, and DNA ligase III is a novel molecular 'nick-sensor' in vitro. Nucleic Acids Res 1996; 24(22):4387-4394.
107. Fornace Jr AJ, Zmudzka B, Hollander MC et al. Induction of β-polymerase mRNA by DNA damaging agents in Chinese hamster ovary cells. Mol Cell Biol 1989; 9:851-853.
108. Chen KH, Yakes FM, Srivastava DK et al. Up-regulation of base excision repair correlates with enhanced protection against a DNA damaging agent in mouse cell lines. Nucleic Acids Res 1998; 26(8):2001-2007.
109. Wood RD, Shivji MK. Which DNA polymerases are used for DNA-repair in eukaryotes? Carcinogenesis 1997; 18(4):605-610.
110. Longley MJ, Pierce AJ, Modrich P. DNA polymerase δ is required for human mismatch repair in vitro. J Biol Chem 1997; 272(16):10917-10921.
111. Matsumoto Y, Kim K, Bogenhagen DF. Proliferating cell nuclear antigen-dependent abasic site repair in Xenopus laevis oocytes: An alternative pathway of base excision DNA repair. Mol Cell Biol 1994; 14(9):6187-6197.

112. Liu L, Mo J, Rodriges-Belmonte EM et al. Identification of a fourth subunit of mammalian DNA polymerase δ. J Biol Chem 2000; 275:18739-18744.
113. Podust VN, Chang L-S, Ott R et al. Reconstitution of human DNA polymerase δ using recombinant baculoviruses. J Biol Chem 2002; 277:3994-3901.
114. Lee MYWT, Tan C-K, Downey KM et al. Further studies on calf thymus DNA polymerase δ purified to homogeneity by a new procedure. Biochemistry 1984; 23:1906-1913.
115. Kelman Z. PCNA: Structure, functions and interactions. Oncogene 1997; 14:629-640.
116. Prelich G, Tan CK, Kostura M et al. Functional identity of proliferating cell nuclear antigen and a DNA polymerase-delta auxiliary protein. Nature 1987; 326(6112):517-520.
117. Ng L, Tan CK, Downey KM et al. Enzymologic mechanism of calf thymus DNA polymerase-δ. J Biol Chem 1991; 266(18):11699-11704.
118. McConnell M, Miller H, Mozzherin DJ et al. The mammalian DNA polymerase δ-proliferating cell nuclear antigen-template-primer complex: Molecular characterization by direct binding. Biochemistry 1996; 35(25):8268-8274.
119. Kong XP, Onrust R, O'Donnell M et al. 3-Dimensional structure of the β-subunit of Escherichia coli DNA polymerase-III holoenzyme—A sliding DNA clamp. Cell 1992; 69(3):425-437.
120. Tomkinson AE, Levin DS. Mammalian DNA ligases. Bioessays 1997; 19(10):893-901.
121. Tomkinson AE, Mackey ZB. Structure and function of mammalian DNA ligases. Mutat Res 1998; 407(1):1-9.
122. Waga S, Stillman B. DNA replication fork in eukaryotic cells. Annu Rev Biochem 1998; 67:721-751.
123. Montecucco A, Rossi R, Levin DS et al. DNA ligase I is recruited to sites of DNA replication by an interaction with prolifirating cell nuclear antigen: Identification of a common targeting mechanism for the assembly of replication factories. EMBO J 1998; 17:3786-3795.
124. Levin DS, Bai W, Yao N et al. An interaction between DNA ligase I and proliferating cell nuclear antigen: Implication for Okazaki fragment synthesis and joining. Proc Natl Acad Sci USA 1997; 94:12863-12868.
125. Soderhall S. DNA ligases during rat liver regeneration. Nature 1976; 260:640-642.
126. Aboussekhra A, Biggerstaff M, Shivji MK et al. Mammalian DNA nucleotide excision repair reconstituted with purified protein components. Cell 1995; 80:859-868.
127. Levin DS, McKenna AE, Motycka TA et al. Interaction between PCNA and DNA ligase I is critical for joining of Okazaki fragments and long-patch base-excision repair. Current Biol 2000; 10(15):919-922.
128. Barnes DE, Tomkinson AE, Lehmann AR et al. Mutations in the DNA ligase I gene of an individual with immunodeficiencies and cellular hypersensitivity. Cell 1992; 69:495-504.
129. Webster ADB, Barnes DE, Arlett CF et al. Growth retardation and immunodeficiency in a patient with mutations in the DNA ligase I gene. Lancet 1992; 339:1508-1509.
130. Prigent C, Satoh MS, Daly G et al. Aberrant DNA repair and DNA replication due to an inherited enzymatic defect in human DNA ligase I. Mol Cell Biol 1994; 14(1):310-317.
131. Harrison C, Ketchen A-M, Redhead NJ et al. Replicatuion failure, genome instability, and increased cancer susceptibility in mice with a point mutation in the DNA ligase I gene. Cancer Res 2002; 62:4065-4074.
132. Bentley DJ, Selfridge J, Millar JK et al. DNA ligase I is required for fetal liver erythropoiesis but is not essentiall for mammalian cells. Nature Genetics 1996; 13:489-491.
133. Bentley DJ, Harrison C, Ketchen A-M et al. DNA ligase I null mouse cells show normal DNA repair activity but altered DNA replication and reduced genome stability. J Cell Sci 2002; 115:1551-1561.
134. Tom S, Henricksen LA, Park MS et al. DNA ligase I and proliferating cell nuclear antigen form a functional complex. J Biol Chem 2001; 276(27):24817-25.
135. Ranalli TA, DeMott MS, Bambara RA. Mechanism underlying replication protein A stimulation of DNA ligase I. J Biol Chem 2002; 277(3):1719-1727.
136. Dianov GL, Jensen BR, Kenny MK et al. Replication protein A stimulates proliferating cell nuclear antigen-dependent repair of abasic sites in DNA by human cell extracts. Biochemistry 1999; 38(34):11021-11025.
137. DeMott M, Zigman S, Bambara RA. Replication protein A stimulates long patch DNA base excision repair. J Biol Chem 1998; 273(42):27492-27498.

138. Mackey ZB, Ramos W, Levin DS et al. An alternative splicing event which occurs in mouse pachytene spermatocytes generates a form of DNA ligase III with distinct biochemical properties that may function in meiotic recombination. Mol Cell Biol 1997; 17(2):989-998.
139. Tomkinson AE, Chen L, Dong Z et al. Completion of base excision repair by mammalian DNA ligases. Prog Nucleic Acid Res Mol Biol 2001; 68:151-164.
140. Taylor RM, Whitehouse J, Cappelli E et al. Role of the DNA ligase III zinc finger in polynucleotide binding and ligation. Nucleic Acids Res 1998; 26(21):4804-4810.
141. Wei YF, Robins P, Carter K et al. Molecular cloning and expression of human cDNAs encoding a novel DNA ligase IV and DNA ligase III, an enzyme active in DNA repair and recombination. Mol Cell Biol 1995; 15(6):3206-3216.
142. Taylor RM, Whitehouse CJ, Caldecott KW. The DNA ligase III zinc finger stimulates binding to DNA secondary structure and promotes end joining. Nucleic Acids Res 2000; 28:3558-3563.
143. Mackey ZB, Niedergang C, Murcia JM et al. DNA ligase III is recruited to DNA strand breaks by a zinc finger motif homologous to that of poly(ADP-ribose) polymerase. Identification of two functionally distinct DNA binding regions within DNA ligase III. J Biol Chem 1999; 274(31):21679-21687.
144. Caldecott KW, McKeown CK, Tucker JD et al. An interaction between the mammalian DNA repair protein XRCC1 and DNA ligase III. Mol Cell Biol 1994; 14:68-76.
145. Caldecott KW, Tucker JD, Stanker LH et al. Characterization of the XRCC1-DNA ligase III complex in vitro and its absence from mutant hamster cells. Nucleic Acids Res 1995; 23(23):4836-4843.
146. Nash RA, Caldecott KW, Barnes DE et al. XRCC1 protein interacts with one of two distinct forms of DNA ligase III. Biochemistry 1997; 36(17):5207-5211.
147. Ljungquist S, Kenne K, Olsson L et al. Altered DNA ligase III activity in the CHO EM9 mutant. Mutat Res 1994; 314(2):177-186.
148. Cappelli E, Taylor R, Cevasco M et al. Involvement of XRCC1 and DNA ligase III gene products in DNA base excision repair. J Biol Chem 1997; 272(38):23970-23975.
149. Caldecott KW, Tucker JD, Thompson LH. Construction of human XRCC1 minigenes that fully correct the CHO DNA repair mutant EM9. Nucleic Acids Res 1992; 20(17):4575-4579.
150. Whitehouse CJ, Taylor RM, Thistlethwaite A et al. XRCC1 stimulates human polynucleotide kinase activity at damaged DNA termini and accelerates DNA single-strand break repair. Cell 2001; 104:107-117.
151. Masson M, Niedergang C, Schreiber V et al. XRCC1 is specifically associated with poly(ADP-ribose) polymerase and negatively regulates its activity following DNA damage. Mol Cell Biol 1998; 18(6):3563-3571.
152. Marintchev A, Mullen MA, Maciejewski MW et al. Solution structure of the single-strand break repair protein XRCC1 N- terminal domain. Nat Struct Biol 1999; 6(9):884-893.
153. Taylor RM, Wickstead B, Cronin S et al. Role of a BRCT domain in the interaction of DNA ligase III-alpha with the DNA repair protein XRCC1. Curr Biol 1998; 8(15):877-880.
154. Dulic A, Bates PA, Zhang X et al. BRCT domain interactions in the heterodimeric DNA repair protein XRCC1-DNA ligase III. Biochemistry 2001; 40(20):5906-5913.
155. Moore DJ, Taylor RM, Clements P et al. Mutation of a BRCT domain selectively disrupts DNA single-strand break repair in noncycling Chinese hamster ovary cells. Proc Natl Acad Sci USA 2000; 97(25):13649-13654.
156. Taylor RM, Moore DJ, Whitehouse J et al. A cell cycle-specific requirement for the XRCC1 BRCT II domain during mammalian DNA strand break repair. Mol Cell Biol 2000; 20(2):735-740.
157. Taylor RM, Thistlethwaite A, Caldecott KW. Central role for the XRCC1 BRCT I domain in mammalian DNA single -strand break repair. Mol Cell Biol 2002; 22:2556-2563.
158. Vidal AE, Boiteux S, Hickson ID et al. XRCC1 coordinates the initial and late stages of DNA abasic site repair through protein-protein interactions. EMBO J 2001; 20(22):6530-6539.
159. Thompson LH, West MG. XRCC1 keeps DNA from getting stranded. Mutat Res 2000; 459(1):1-18.
160. Harrington JJ, Leiber MR. The characterization of a mammalian DNA structure specific endonuclease. EMBO J 1994; 13(5):1235-1246.

161. Shibata Y, Nakamura T. Defective flap endonuclease 1 activity in mammalian cells is associated with impaired DNA repair and prolonged S phase delay. J Biol Chem 2002; 277(1):746-754.
162. Qiu J, Li X, Frank G et al. Cell cycle-dependent and DNA damage-inducible nuclear localization of FEN-1 nuclease is consistent with its dual functions in DNA replication and repair. J Biol Chem 2001; 276(7):4901-4908.
163. Murray JM, Tavassoli M, Al-Harithy R et al. Structural and functional conservation of the human homolog of the Schizosaccharomyces pombe rad2 gene, which is required for chromosome segregation and recovery from DNA damage. Mol Cell Biol 1994; 14:4878-4888.
164. Reagan MS, Pittenger C, Siede W et al. Characterization of a mutant strain of Saccharomyces cerevisiae with a deletion of the RAD27 gene, a structural homolog of the RAD2 nucleotide excision repair gene. J Bacteriol 1995; 177:364-371.
165. Matsuzaki Y, Adachi N, Koyama H. Vertebrate cells lacking FEN-1 endonuclease are viable but hypersensitive to methylating agents and H_2O_2. Nucleic Acids Res 2002; 30(14):3273-3277.
166. Prasad R, Dianov GL, Bohr VA et al. FEN1 stimulation of DNA polymerase β mediates an excision step in mammalian long patch base excision repair. J Biol Chem 2000; 275:4460-4465.
167. DeMott MS, Shen B, Park M et al. Human RAD2 homolog 1 5'- to 3'-Exo/Endonuclease can efficiently excise a displaced DNA fragment containing a 5'-terminal abasic lesion by endonuclease activity. J Biol Chem 1996; 47:30068-30076.
168. Li X, Li J, Harrington J et al. Lagging strand DNA synthesis at the eukaryotic replication fork involves binding and stimulation of FEN-1 by proliferating cell nuclear antigen. J Biol Chem 1995; 270(38):22109-22112.
169. Wu X, Li J, Li X et al. Processing of branched DNA intermediates by a complex of human FEN-1 and PCNA. Nucleic Acids Res 1996; 24(11):2036-2043.
170. Gary R, Kim K, Cornelius HL et al. Proliferating cell nuclear antigen facilitates excision in long-patch base excision repair. J Biol Chem 1999; 7:4354-4363.
171. Dianova II, Bohr VA, Dianov GL. Interaction of human AP endonuclease 1 with flap endonuclease 1 and proliferating cell nuclear antigen involved in long-patch base excision repair. Biochemistry 2001; 40(42):12639-44.
172. Brosh Jr RM, von Kobbe C, Sommers JA et al. Werner syndrome protein interacts with human flap endonuclease 1 and stimulates its cleavage activity. EMBO J 2001; 20(20):5791-5801.
173. Wang ZQ, Stingl L, Morrison C et al. PARP is important for genomic stability but dispensable in apoptosis. Genes & Development 1997; 11(18):2347-2358.
174. Menissier-de Murcia J, Niedergang C, Trucco C et al. Requirement of poly(ADP-ribose) polymerase in recovery from DNA damage in mice and in cells. Proc Natl Acad Sci USA 1997; 94:7303-7307.
175. Trucco C, Oliver FJ, de Murcia G et al. DNA repair defect in poly(ADP-ribose) polymerase-deficient cell lines. Nucleic Acids Res 1998; 26(11):2644-2649.
176. Satoh MS, Lindahl T. Role of poly(ADP-ribose) formation in DNA repair. Nature 1992; 356(6367):356-358.
177. Satoh MS, Poirier GG, Lindahl T. NAD^+-dependent repair of damaged DNA by human cell extracts. J Biol Chem 1993; 268(8):5480-5487.
178. D'Amours D, Desnoyers S, D'Silva I et al. Poly(ADP-ribosyl)ation reactions in the regulation of nuclear functions. Biochem J 1999; 324:249-268.
179. Vodenicharov MD, Sallmann FR, Satoh MS et al. Base excision repair is efficient in cells lacking poly(ADP-ribose) polymerase 1. Nucleic Acids Res 2000; 28:3887-3896.
180. Prasad R, Lavrik OI, Kim SJ et al. DNA polymerase beta-mediated long patch base excision repair. Poly(ADP-ribose)polymerase-1 stimulates strand displacement DNA synthesis. J Biol Chem 2001; 276(35):32411-32414.
181. Dantzer F, de la Rubia G, Murcia JMD et al. Base excision repair is impaired in mammalian cells lacking poly(ADP-ribose) polymerase-1. Biochemistry 2000; 39(25):7559-7569.
182. Wilson SH, Kunkel TA. Passing the baton in base excision repair. Nature Struct Biol 2000; 7:176-178.
183. Hill JW, Hazra TK, Izumi T et al. Stimulation of human 8-oxoguanine-DNA glycosylase by AP-endonuclease: Potential coordination of the initial steps in base excision repair. Nucleic Acids Res 2001; 29:430-438.

184. Vidal AE, Hickson ID, Boiteux S et al. Mechanism of stimulation of the DNA glycosylase activity of hOGG1 by the major human AP endonuclease: Bypass of the AP lyase activity step. Nucleic Acids Res 2001; 29:1285-1292.
185. Walker LJ, Craig RB, Harris AL et al. A role for the human DNA repair enzyme HAP1 in cellular protection against DNA damaging agents and hypoxic stress. Nucleic Acids Res 1994; 22:4884-4889.
186. Bennett RAO, Wilson 3rd DM, Wong D et al. Interaction of human apurinic endonuclease and DNA polymerase beta in the base excision repair pathway. Proc Natl Acad Sci USA 1997; 94(14):7166-7169.
187. Dimitriadis EK, Prasad R, Vaske MK et al. Thermodynamics of human DNA ligase I trimerization and association with DNA polymerase beta. J Biol Chem 1998; 273(32):20540-20550.
188. McCullough AK, Dodson ML, Lloyd RS. Initiation of Base Excision Repair: Glycosylase Mechanism and Structure. Annu Rev Biochem 1999; 68:255-285.
189. Dianov G, Bischoff C, Piotrowski J et al. Repair pathways for processing of 8-oxoguanine in DNA by mammalian cell extracts. J Biol Chem 1998; 273:33811-33816.
190. Dianov GL, Thybo T, Dianova II et al. Single nucleotide patch base excision repair is the major pathway for removal of thymine glycol from DNA in human cell extracts. J Biol Chem 2000; 275(16):11809-11813.
191. Fortini P, Parlanti E, Sidorkina OM et al. The type of DNA glycosylase determines the base excision repair pathway in mammalian cells. J Biol Chem 1999; 274:15230-15236.
192. Waters TR, Gallinari P, Jiricny J et al. Human thymine DNA glycosylase binds to apurinic sites in DNA but is displaced by human apurinic endonuclease 1. J Biol Chem 1999; 274(1):67-74.
193. Nakamura J, La DK, Swenberg JA. 5-nicked apurinic/apyrimidinic sites are resistant to beta-elimination by beta-polymerase and are persistent in human cultured cells after oxidative stress. J Biol Chem 2000; 275(8):5323-5328.
194. Podlutsky AJ, Dianova II, Podust VN et al. Human DNA polymerase beta initiates DNA synthesis during long- patch repair of reduced AP sites in DNA. EMBO J 2001; 20(6):1477-1482.
195. Kim K, Biade S, Matsumoto Y. Involvement of flap endonuclease 1 in base excision DNA repair. J Biol Chem 1998; 273(15):8842-8848.
196. Frosina G, Fortini P, Rossi O et al. Two pathways for base excision repair in mammalian cells. J Biol Chem 1996; 271(16):9573-9578.
197. Dianov GL, Prasad R, Wilson SH et al. Role of DNA polymerase β in the excision step of long patch mammalian base excision repair. J Biol Chem 1999; 274(20):13741-13743.
198. Warbrick E. PCNA binding through a conserved motif. Bioessays 1998; 20(3):195-199.
199. Tom S, Henricksen LA, Bambara RA. Mechanism whereby proliferating cell nuclear antigen stimulates flap endonuclease1. J Biol Chem 2000; 275:10498-10505.
200. Lu XY, Tan C-K, Zhou J-Q et al. Direct interaction of Proliferating Cell Nuclear Antigen with the small subunit of DNA polymerase δ. J Biol Chem 2002; 277:24340-24345.
201. Olsen LC, Aasland R, Wittwer CU et al. Molecular cloning of human uracil-DNA glycosylase, a highly conserved DNA repair enzyme. EMBO J 1989; 8:3121-3125.
202. Krokan HE, Otterlei M, Nilsen H et al. Properties and functions of human uracil-DNA glycosylase from the UNG gene. Prog Nucleic Acid Res Mol Biol 2001; 68:365-386.
203. Slupphaug G, Markussen FH, Olsen LC et al. Nuclear and Mitochondrial Forms of Human Uracil-DNA Glycosylase Are Encoded by the Same Gene. Nucleic Acids Res 1993; 21(11):2579-2584.
204. Haushalter KA, Stukenberg PT, Kirschner MW et al. Identification of a new uracil-DNA glycosylase family by expression cloning using synthetic inhibitors. Curr Biol 1999; 9(4):174-185.
205. Boorstein RJ, Cummings Jr A, Marenstein DR et al. Definitive identification of mammalian 5-hydroxymethyluracil DNA N-glycosylase activity as SMUG1. J Biol Chem 2001; 276(45):41991-41997.
206. Neddermann P, Gallinari P, Lettieri T et al. Cloning and expression of human G/T mismatch-specific thymine-DNA glycosylase. J Biol Chem 1996; 271(22):12767-12774.
207. Hendrich B, Hardeland U, Ng HH et al. The thymine glycosylase MBD4 can bind to the product of deamination at methylated CpG sites. Nature 1999; 401(6750):301-304.
208. Petronzelli F, Riccio A, Markham GD et al. Biphasic kinetics of the human DNA repair protein MED1 (MBD4), a mismatch-specific DNA N-glycosylase. J Biol Chem 2000; 275(42):32422-32429.

209. Slupska MM, Baikalov C, Luther WM et al. Cloning and sequencing a human homolog (hMYH) of the Escherichia coli mutY gene whose function is required for the repair of oxidative DNA damage. J Bacteriol 1996; 178(13):3885-3892.
210. Slupska MM, Luther WM, Chiang JH et al. Functional expression of hMYH, a human homolog of the Escherichia coli MutY protein. J Bacteriol 1999; 181(19):6210-6213.
211. Aspinwall R, Rothwell DG, Roldan-Arjona T et al. Cloning and characterization of a functional human homolog of Escherichia coli endonuclease III. Proc Natl Acad Sci USA 1997; 94(1):109-114.
212. Ikeda S, Biswas T, Roy R et al. Purification and characterization of human NTH1, a homolog of Escherichia coli endonuclease III. Direct identification of Lys-212 as the active nucleophilic residue. J Biol Chem 1998; 273(34):21585-21593.
213. Sarker AH, Ikeda S, Nakano H et al. Cloning and characterization of a mouse homologue (mNthl1) of Escherichia coli endonuclease III. J Mol Biol 1998; 282(4):761-774.
214. Rosenquist TA, Zharkov DO, Grollman AP. Cloning and characterization of a mammalian 8-oxoguanine DNA glycosylase. Proc Natl Acad Sci USA 1997; 94:7429-7434.
215. Radicella JP, Dherin C, Desmaze et al. Cloning and characterization of hOGG1, a human homolog of the OGG1 gene of Saccharomyces cerevisiae. Proc Natl Acad Sci USA 1997; 94:8010-8015.
216. Roldan-Arjona T, Wei YF, Carter KD et al. Molecular cloning and functional expression of a human cDNA encoding the antimutator enzyme 8-hydroxyguanine-DNA glycosylase. Proc Natl Acad Sci USA 1997; 94:8016-8020.
217. Samson L, Derfler B, Boosalis M et al. Cloning and characterization of a 3-methyladenine DNA glycosylase cDNA from human cells whose gene maps to chromosome-16. Proc Natl Acad Sci USA 1991; 88:9127-9131.
218. Chakravarti D, Ibeanu GC, Tano K et al. Cloning and expression in Escherichia coli of a human cDNA encoding the DNA-repair protein N-methylpurine-DNA glycosylase. J Biol Chem 1991; 266:15710-15715.
219. Allinson SL, Dianova II, Dianov GL. Poly(ADP-ribose) polymerase in base excision repair: Always engaged, but not required for DNA damage processiong. Acta Biochim Pol 2003; 50:169-179.
220. Dianov GL, Sleeth KM, Dianova II et al. Repair of abasic sites in DNA by mammalian cells. Mutation Res. In Press.

CHAPTER 3

Detection and Repair of DNA Single-Strand Breaks

Keith W. Caldecott

Abstract

DNA single-strand breaks can arise indirectly, as normal intermediates of DNA base excision repair, or directly from damage to deoxyribose. Because single-strand breaks are induced by endogenous reactive molecules such as reactive oxygen species, these lesions pose a continuous threat to genetic integrity. XRCC1 protein plays a major role in facilitating the repair of single-strand breaks in mammalian cells, via an ability to interact with multiple enzymatic components of repair reactions. Here, the protein-protein interactions facilitated by XRCC1, and the repair processes in which these interactions operate, are reviewed. Models for the repair of single-strand breaks during base excision repair and at direct breaks are presented.

The Origin and Structure of DNA Single-Strand Breaks

Single-strand breaks (SSBs) are discontinuities in the sugar-phosphate backbone of one strand of a single strand of a DNA duplex. Hundreds of thousands of cellular SSBs arise from DNA damage each day, which if un-repaired can be converted into potentially clastogenic and/or lethal DNA double-strand breaks. The two major sources of endogenous SSBs are sugar damage and DNA base damage arising primarily from attack by reactive oxygen species (ROS) and other electrophilic molecules, and from the intrinsic instability of DNA.[1,2] SSBs can also be induced by exposure to environmental genotoxins, ultraviolet light, or ionizing radiation. Whereas SSBs can arise from sugar damage directly, by disintegration of deoxyribose, those from base damage can arise indirectly, through enzymatic excision of the damaged base and abasic site by DNA base excision repair (BER). Another source of SSBs is topoisomerase I (topo1), which nicks and reseals DNA as part of its catalytic activity. Under certain conditions such SSBs can be "uncoupled" from topo1 activity and converted into single stranded or double stranded breaks during transcription or DNA replication. Most SSBs are accompanied by the loss of a single nucleotide at the site of the break and are thus actually single-stranded gaps. In the case of direct SSBs induced by endogenous ROS or IR base loss can arise when the damaged sugar fragments. In the case of indirect SSB, base loss occurs when the damaged base is excised during BER. SSB also typically possess 'damaged' termini that lack the conventional 5'-phosphate or 3'-hydroxyl moieties. The 3'-termini of direct SSB induced by endogenous ROS or IR possess primarily monophosphate (~70%) or phosphoglycolate (~30%) end groups.[2] Although most 5'-termini at such breaks are of the conventional 5'-phosphate format, some

Eukaryotic DNA Damage Surveillance and Repair, edited by Keith W. Caldecott.

may possess a hydroxyl end group. Topo1-induced breaks possess both 5'-hydroxyl and 3'-phosphate damaged termini i.e. the reverse format to that present at conventional DNA ends.[3] Indirect SSB arising during BER possess are of two formats, depending on the mechanism of base excision, a normal 3'-terminus and a 5'-deoxyribose phosphate (dRP) terminus, or a normal 5'-terminus and a 3'-α,β unsaturated aldehyde (αβ) terminus.

The Threat of Single-Strand Breaks to Genetic Stability and Cell Death

It is generally accepted that SSBs are far less toxic to cells than their double-stranded counterpart. However, this largely reflects the ability of cells to rapidly repair large numbers of SSBs, rather than an intrinsic lack of toxicity of these lesions. Indeed, any SSBs not repaired prior to DNA replication or actively "tolerated" in the appropriate way during DNA replication will become DSBs. Given the frequency of "spontaneous" SSBs, it is thus not surprising that cells have evolved highly efficient mechanisms to minimize their impact. Perhaps the most compelling experimental evidence for this emerges from mammalian cells in which there is a defect in SSBR. Rodent cells lacking the molecular scaffold protein XRCC1[4] exhibit elevated cellular sensitivity to a range of agents that induce the SSBs directly or indirectly, including alkylating agents, ionizing radiation, hydrogen peroxide, and camptothecin.[5] Moreover, XRCC1$^{-/-}$ embryos fail to develop beyond day 7.5/8.5, suggesting that endogenous levels of SSBs are incompatible with embryonic development in the absence of efficient SSBR.[6] Un-rejoined SSBs also threaten genetic stability. *XRCC1*-mutant rodent cells exhibit increased frequencies of "spontaneous" and induced chromosome aberrations, including translocations, and genetic deletion.[5,7] The simplest explanation for this is that un-rejoined SSBs are converted into DSBs during chromosomal replication and thereby increase the risk of unwanted genetic rearrangement. In addition to gross genetic changes SSBs can also result in point mutations, due presumably to occasional DNA polymerase mis-incorporation during chromosome replication. Finally, SSBs may also affect transcription. A single nucleotide gap in the template strand fails to stop elongation by T7 RNA polymerase but rather results in a transcript shortened by one nucleotide, in vitro.[8] In contrast, SSBs possessing 'damaged' 3'-phosphate termini can block elongation by T7 RNA polymerase, due apparently to effects of charge repulsion between the neighboring 3'- and 5'-terminal phosphates. Thus, whereas SSBs with normal termini may result in mutant transcripts, SSBs with 3'-phosphate termini, such as those created by endogenous ROS or IR, may block transcription.

Polypeptides Involved in DNA Single-Strand Break Repair

Polypeptides operative during SSBR are discussed below, and a model for SSBR based on our current understanding of these polypeptides is presented (Fig. 1).

PARP-1, XRCC1 and the Detection of SSBs

PARP-1 is an abundant nuclear protein that can function as a molecular nick sensor, and is important for genetic stability and for cellular resistance to ionising radiation and alkylating agents.[9-13] These phenotypes are consistent with a direct role for PARP-1 in DNA repair. However, indirect roles cannot be ruled out because PARP-1 appears to strongly influence chromatin architecture and gene expression.[14-20] In addition, PARP-1 is also implicated in the induction of cell death.[21,22] While at low levels of genotoxic damage PARP-1 appears to promote survival, at lethal levels of damage PARP-1 triggers cell death by promoting nuclear translocation of apoptosis-inducing factor (AIF), most likely via excessive depletion of cellular NAD+.

PARP-1 can covalently ribosylate both itself and, to a lesser extent, a limited number of other acceptor proteins with long negatively charged chains of poly (ADP-ribose). The poly-

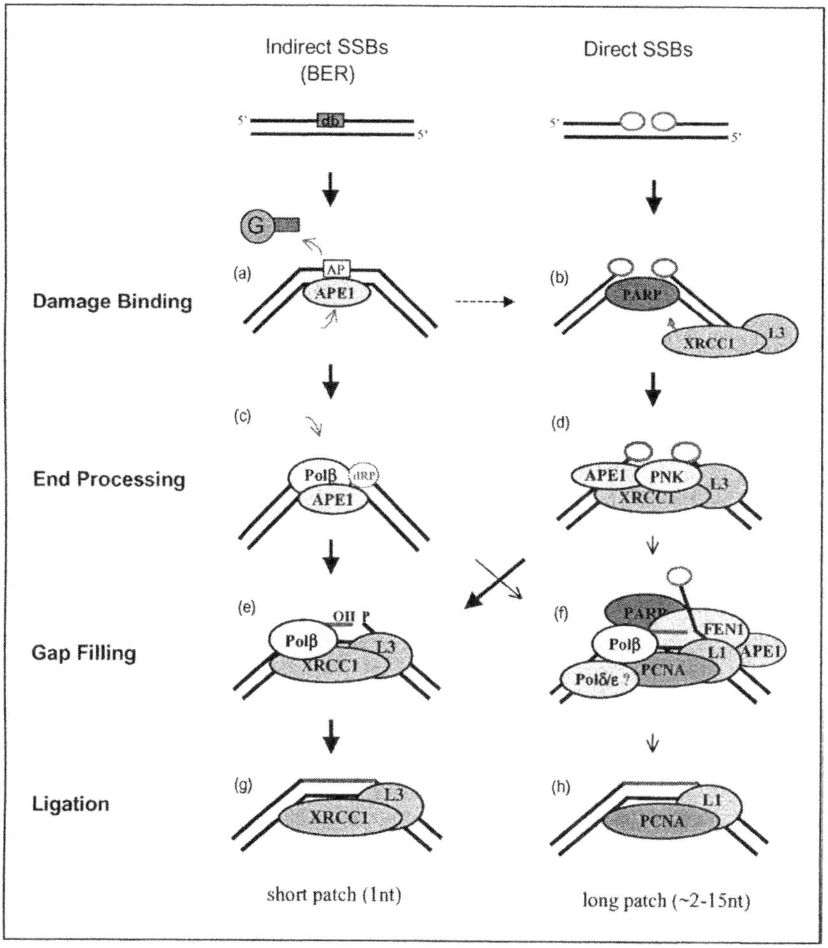

Figure 1. Model for Mammalian SSBR at Direct Breaks and at Indirect Breaks During BER. At indirect breaks created during BER, a damaged DNA Base ("db") is removed by a DNA glycosylase ("G"), which is then itself displaced from the abasic site ("AP") by APE1 (a), which then recruits Polβ and cleaves the abasic site (c). Polβ inserts a single nucleotide and repairs the 5'-deoxyribose phosphate terminus ("dRP") created by APE1, (e). This results in a ligatable nick that is sealed by recruitment of XRCC1-Lig3α, (g). Under some circumstances Polβ cannot remove the 5'-dRP terminus (e.g., if the dRP is oxidised or reduced). At these 'stalled' repair events, recruitment of PARP, PCNA, and FEN1 stimulates extension of the gap by 2-15 nucleotides and cleavage of the resulting flap, (f). The nick is subsequently ligated by Lig1, (h). Note that Polδ/ε may participate in long patch repair under some circumstances. (b), Direct SSBs arising from sugar damage, and possibly a subset of SSBs arising during BER [e.g., abasic sites that become 'uncoupled' from APE1 or that are cleaved by AP lyase (dotted arrow)], are bound by PARP-1 or PARP-2 which are then activated and automodified. Automodified PARP recruits XRCC1-Lig3α to the SSB, which replaces PARP and forms a molecular scaffold. (d), Damaged 5'- or 3'-termini (red circles) are converted to 5'-phosphate and 3'-hydroxyl moieties by APE1 or PNK, which interact with and are stimulated by XRCC1. Recruitment of Polβ (e), enables single nucleotide gap filling at most SSBs followed by ligation by Lig3α, (g). As described for BER, above, at some direct breaks, gap filling may be conducted in long patch mode by Polβ or Polδ/ε (f), with ligation conducted by Lig1, (h). Damaged termini are indicated by red circles and nascent DNA from gap filling by a red line. The major pathways are indicated by large arrows and the minor pathways by small arrows. Please view the color image at www.Eurekah.com.

mer is synthesized by successive addition of ADP-ribose units at the expense of NAD^+. PARP-1 activity is stimulated more than 500-fold upon binding of the polypeptide to DNA single or double strand breaks. Recently, another DNA damage-dependent PARP, PARP-2, was identified.[23] PARP-1 appears to function as a homodimer, and possibly as a heterodimer with PARP-2. The ability of PARP-1 to bind rapidly to DNA strand breaks is conferred by two unusual amino-terminal zinc fingers of the type $CX_2CX_{28-30}HX_2C$.[10] An intriguing feature of PARP-1 is the structure of this molecule bound to a DNA nick.[24] When measured by electron microscopy, a 139 base pair nicked duplex exhibits a kinked conformation, with an angle at the apex of the duplex where the nick is located of ~122°. This kinked confirmation is accentuated by binding by PARP-1 to the nick, since the angle at the apex is decreased to ~102°. This appears to be a general feature of DNA intermediates of BER/SSBR (see Fig. 1), since DNA bound by AP endonuclease-1 (APE1) or DNA polymeraseβ (Polβ) is also kinked.[25,26]

The binding of PARP-1 to DNA breaks is very transient because autoribosylation results in dissociation of the polymerase from DNA, most likely by charge repulsion.[27,28] PAR is also rapidly degraded within minutes, by PAR glycohydrolase, further underlining the transient nature of PARP-1 activation. The rapid dissociation of PARP-1 from breaks appears to be important to allow subsequent access to the break by other DNA repair proteins. This is illustrated by the inhibitory effect of PARP-1 on the repair of direct SSBs or indirect SSBs during BER in assays conducted in vitro in the absence of its cofactor, NAD^+, during which PARP-1 can bind to DNA breaks but cannot be activated.[29-32]

Transient binding of PARP-1 to DNA breaks may also serve a signalling role to recruit other DNA repair proteins. This is suggested by the observation that autoribosylated PARP-1,[33] and possibly PARP-2, interact with and recruit the SSBR polypeptide XRCC1.[23,33-36] XRCC1 is a critical component of the SSBR machinery, and is required for this process in response to SSBs arising either indirectly during BER or directly. This is exemplified by the observation that XRCC1-mutant cell lines are hypersensitivity to a broad range of genotoxins, such as the monofunctional alkylating agents EMS and MMS, H_2O_2, camptothecin, ionising radiation, ENU, MNNG, MMC, UVC, UVA, near visible and blue light, and heavy metals.[5,37-41] The rate of rejoining of SSBs induced by such agents is reduced several fold in XRCC1-mutant CHO cells when compared to wild type cells cells.[37,39-42] As will become evident during this chapter, XRCC1 appears to be a molecular scaffold protein that recruits, stabilises, and stimulates other components of the SSBR process. The interactions mediated by XRCC1 are illustrated in Figure 2.

XRCC1 preferentially interacts with autoribosylated PARP-1 and PARP-2, and thus active PARP that has bound a strand break. A possible explanation for the ability of XRCC1 to distinguish between inactive and active PARP emerged with the discovery that the XRCC1 BRCT I domain that binds PARP-1 and PARP-2 contains a degenerate consensus motif for interaction with PAR polymer.[43] Ensuring that XRCC1 preferentially binds activated PARP-1 and PARP-2 can thereby ensure that an XRCC1 protein scaffold is assembled specifically at sites of strand breakage.

It should be noted that DNA repair reactions conducted by cell free extracts have so far failed to observe a requirement for PARP-1 for efficient repair of single-strand breaks induced by ionising radiation, restriction endonucleases, or alkylating agents.[30-32,44,45] This may reflect that the active recruitment of XRCC1 to breaks is not required for repair in simple cell-free reactions, perhaps because the high concentration of DNA breaks in such experiments prevent their detection being a rate-limiting step. Active recruitment of XRCC1 and its protein partners may only be important to accelerate the detection and repair of SSBs within the context of the nuclear genome, in which DNA breaks are relatively rare and distributed amongst many kilobases of undamaged DNA.

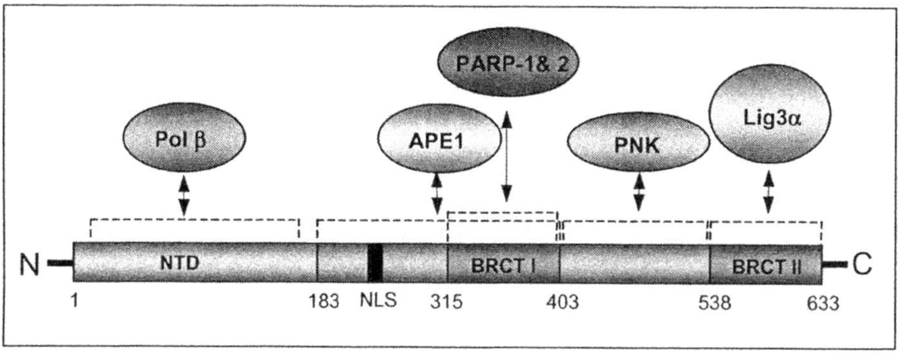

Figure 2. Protein-Protein Interactions Mediated by Human XRCC1. NTD, N-terminal domain; NLS, nuclear localisation signal. Dotted lines denote the domains within which the indicated interactions occur.

The notion that PARP-1 might rapidly recruit XRCC1 to direct DNA breaks is attractive, because direct breaks arise in an "unscheduled" manner. Such breaks are presumably free in solution and vulnerable to unwanted rearrangement or degradation, and therefore require rapid detection.[4] It is more difficult to envisage a need for PARP to detect breaks during early stages of BER, since current models for this process suggest that breaks created by APE1 can be passed directly to DNA polymerase β[25,26,46] which in turn may itself then recruit XRCC1/DNA ligase IIIα for the completion of DNA repair[47] (Fig. 1a,c,e,g). There is evidence both for[48,49] and against[32] a role for PARP-1 in BER. One possible role could be to recruit XRCC1 to SSBs at a fraction of abasic sites that become "uncoupled" from APE1 after cleavage or which are cleaved independently of APE1, such as by an AP lyase activity (Fig. 1a, dotted arrow). Alternatively, PARP-1 may play a role during a later stage of BER, by promoting long patch gap filling by DNA polymerase-β and flap endonuclease-1 (FEN1), for example[30,36,49,50] (Fig. 1f).

AP Endonuclease-1 (APE1)

APE1 is the major AP endonuclease activity in mammalian cells and cleaves the sugar phosphate backbone immediately 5' of abasic sites.[51] Abasic sites arise by spontaneous base loss or as intermediates during the BER of modified bases. In addition, modified abasic sites can arise from oxidative damage to deoxyribose, some of which are still substrates for APE1.[52,53] APE1 is thus the major source of indirect SSBs arising during BER. At abasic sites created by DNA glycosylase during BER, recruitment of APE1 may displace the DNA glycosylase from the abasic site (Fig.1a) and then both recruit DNA polymerase-β (Polβ) and cleave the abasic site (Fig. 1c).[25,26,46] This ordered sequence of step-wise interactions has prompted a model in which an enzyme involved in one step of the reaction receives its DNA substrate as an enzyme-product complex from the previous step.[25,26,51] In this way, repair intermediates can be handed from one enzyme to the next in a molecular relay that is driven by sequential protein-protein and protein-DNA interactions. The repair of direct SSBs may be coordinated in a similar fashion.[4,54]

The termini of strand breaks are often of non-conventional chemistry and require enzymatic processing to convert them to the 3'-hydroxyl and 5'-phosphate moieties necessary for gap filling and DNA ligation. In addition to its ability to cleave abasic sites APE1 is also able to remove, albeit to a lesser extent, 3'-phosphoglycolate and 3'-phosphate from direct breaks and 3'-unsaturated aldehydes that result from cleavage of abasic sites by an AP lyase activity[55-58] (Fig. 1d). During BER, some oxidised or reduced 5'-cleaved abasic sites are resistant to removal by the normal mechanism, which employs the dRPase activity of Polβ, thereby requiring long patch gap filling. It has been suggested that under these circumstances, APE1 may help recruit and/or stimulate PCNA, FEN1, and DNA ligase I activities at sites of stalled BER, helping circumvent the blockage by promoting long patch BER[59-61] (Fig. 1f).

Recently, a yeast two-hybrid screen revealed that APE1 binds to XRCC1 by interacting with the central section of the latter protein (Fig. 2).[62] XRCC1 binds to the amino terminus of APE1, and can weakly stimulate (~3-fold) APE1 endonuclease activity if the latter is present in limiting amounts. This raises the possibility that XRCC1 may influence the creation of a SSB during BER in addition to the subsequent steps of removing the SSB. However, XRCC1 also weakly stimulated the 3'-end processing activity of APE1, which would be consistent with its identified role in processing 3'- and 5'-ends with human polynucleotide kinase (Fig. 1d and see below). More recently, APE1 was also reported to possess a 3'-exonuclease activity that can remove mismatched nucleotides incorporated by Polβ.[63] Whether of not XRCC1 stimulates this activity is not known.

Polynucleotide Kinase (PNK)

Human polynucleotide kinase (PNK), a homologue of T4 PNK, possesses both 5'-DNA kinase activity and 3'-phosphatase activity.[64,65] T4 PNK is involved in RNA repair, operating to counteract a host-cell defence mechanism that degrades tRNA. Unlike T4 PNK, the eukaryotic enzyme prefers gapped or nicked DNA as a substrate, prompting the suggestion that human PNK might process DNA termini during DNA repair[65,66] (Fig. 1d). 3'-phosphate termini are present at many direct SSBs arising from sugar damage by oxidising agents and free radicals. Another common type of 3'-terminus at direct SSBs is 3'-phosphoglycolate, and this too can become 3'-phosphate if the enzyme tyrosyl phosphodiesterase-1 (TDP1) first removes the glycolate.[67] Finally, 5'-hydroxyl and 3'-phosphate termini are also created at nicks that result from abortive topoisomerase I activity, such as that induced by the anticancer drug camptothecin.

Evidence that mammalian PNK functions in DNA repair emerged with the recovery of human PNK from a yeast two-hybrid screen with XRCC1.[4] XRCC1 stimulates both the 5'-kinase and 3'-phosphatase activities of PNK. Consistent with this, PNK is rate limiting for the repair of SSBs with 3'-phosphate and 5'-hydroxyl termini in repair reactions driven by cell extracts that lack XRCC1. Further evidence that eukaryotic PNK in is involved in DNA repair in vivo arose recently with creation of cells lacking Pnk in *S. pombe*, which exhibit hypersensitivity to campothecin and ionising radiation.[68] In addition to stimulating PNK activity, it has been proposed that the interaction with XRCC1 serves to recruit PNK to chromosomal SSBs[4] (Fig. 1d).

DNA Polymerase-β (Polβ)

Polβ is a small, single-subunit, DNA polymerase that is associated primarily with filling small gaps during DNA base excision repair.[69,70] In addition, Polβ is also implicated in gap filling at direct SSBs.[65,71-74] Note, however, that there is extensive redundancy in the gap filling role fulfilled by Polβ, with Polδ/ε seemingly able to compensate efficiently for the absence of this polymerase, albeit with delayed kinetics.[75-79] Polβ has been characterised extensively at the biochemical and structural level, and is comprised of an 8 kDa amino terminal domain and a C-terminal 31 kDa domain.[69] The 8 kDa amino terminal domain possesses dRP lyase activity and thereby removes from the 5'-termini of a SSB the 5'-deoxyribose phosphate created by AP endonuclease during BER.[80] The C-terminal domain contains the DNA polymerase active site.

There are two sub-pathways for BER, and most likely for the repair of direct SSBs too (see references above). These are short patch repair, involving incorporation of a single nucleotide (Fig. 1a,c,e,g), and long patch repair, involving insertion of 2-10 nucleotides and consequently single-strand displacement (Fig. 1a,c,f,h). Polβ appears to be the primary polymerase employed for both processes, in reactions conducted by cell extracts at least, though the PCNA-dependent polymerases polδ/ε can also conduct gap filling and may be involved in long patch repair under some circumstances. It is not known how the choice between Polβ subpathways is made in cells, but cell-free reactions suggest that the nature of the SSB is one key factor.[52,81,82] In cell free extracts, Polβ appears to conduct short patch gap filling if the 5'-terminus of the break is a simple deoxyribose phosphate (dRP), but long patch filling if the dRP is reduced or oxidised and thereby a poor substrate for the dRP lyase activity of Polβ (Fig. 1c). Under these circumstances, Polβ may stall and then, possibly with the help of PARP-1 and FEN-1 (Fig. 1f), continue DNA synthesis to displace the blocked 5'-terminus as a single-strand flap that can be removed by FEN-1,[50,83] which in turn may be stimulated by PCNA.[81,84] To date, evidence from cell free reactions best fit a model in which short patch repair is completed by XRCC1-Lig3 (Fig. 1g) and long patch repair by Lig1 (Fig. 1h).[74,85,86]

In 1996 Polβ was identified as a partner of XRCC1.[34,87] This interaction occurs within the palm-thumb domains of the DNA polymerase active site and an amino terminal domain (NTD) of XRCC1[88-90] (Fig. 2). The structures of Polβ /gapped DNA and of the XRCC1 NTD have been solved by crystallography and NMR, respectively. A combination of biochemical experiments and molecular modelling based on these structures has suggested that the NTD may be able to bind both DNA at the site of a single-stranded gap and the palm-thumb domains of Polβ. In this way, the XRCC1- Polβ complex might surround a DNA molecule at the site of a SSB, perhaps protecting the undamaged strand from nucleolytic attack, which might otherwise create a DSB.[88,91] This interaction could also help recruit XRCC1 to a Polβ-bound SSB (Fig. 1c,e) or conversely Polβ to an XRCC1-bound SSB (Fig. 1d,e), depending on which pathway is in use.

DNA Ligase III

Although involved in the last step of repair, DNA ligase III was the first XRCC1 partner to be discovered.[92,93] The DNA ligase III gene encodes at least two nuclear isoforms, denoted DNA ligase IIIα and DNA ligase IIIβ, which differ in their C-termini.[94-96] The interaction between Lig3α and XRCC1 is mediated by their C-termini, both of which encode a BRCT domain[95,97] (Fig. 2). In human Lig3β, the 77 C-terminal amino acids that encode the BRCT domain in Lig3α are replaced by 17 different amino acids, and consequently Lig3β does not interact with XRCC1. The XRCC1 C-terminal BRCT domain was the first such structure to be solved and has provided a paradigm for BRCT domains in other proteins.[98] Most of the cellular complement of Lig3α appears to be constitutively bound to XRCC1.

Levels of Lig3α polypeptide and activity are reduced ~5-fold in cells lacking XRCC1,[92,93,99,100] and inhibitors of the proteosome increase Lig3α levels in such cells.[101] XRCC1 thus appears to be required for the stability of cellular Lig3α, preventing its degradation by the proteosome. In cell-free BER reactions in which the repair of a circular plasmid harbouring a defined abasic site was examined, *XRCC1*-mutant EM-C11 extracts exhibited reduced ability to complete short patch BER[85] but not long patch BER, which appears to utilise Lig1 (see Fig. 1g,h). This defect was complemented by addition of recombinant human Lig3α or T4 DNA ligase, confirming that repair was stalled at the point of ligation. It is likely that, in addition to maintaining the stability of cellular Lig3α, an important function of the interaction with XRCC1 is to target the DNA ligase to sites of strand breakage. In support of this, deletion or mutation of the BRCT domain in XRCC1 that binds Lig3α prevents the ability of XRCC1 to facilitate rapid rejoining of cellular SSBs following exposure to alkylating agents, even if the level of Lig3α is increased with proteosome inhibitors.[101,102]

An intriguing feature of Lig3α and Lig3β is the presence of a putative zinc finger at the amino terminus.[96] This motif exhibits ~ 40% identity with the two amino terminal zinc fingers present in PARP-1. In PARP-1, the zinc fingers confer a "nick sensing" function that is required for this polypeptide to rapidly bind and be activated by DNA single- or double-strand breaks (see above). The zinc finger in Lig3 also has the ability to bind oligonucleotide duplexes possessing simple nicks, though there is conflicting data concerning whether this motif is required for this purpose, in vitro.[103-105] It seems unlikely that the zinc finger will be required for binding and ligation of simple nicks, in vivo, given that all DNA ligases possess an intrinsic nick binding activity within their conserved catalytic domains. Indeed, human Lig3 lacking a zinc finger is able to complement the conditional lethal phenotype of *E. coli* lig mutants.[105]

In contrast to simple nicks, there is considerable evidence that the Lig3 zinc finger confers upon this ligase an ability to bind to DNA with unusual secondary structure, such as nicked RNA homopolymers and DNA duplexes with single-stranded flaps, single-stranded gaps, or Holliday junctions.[103,104] Moreover, the zinc finger appears to stimulate rejoining of double strand breaks located near to these structures. It thus seems likely that the zinc finger increases the substrate specificity of Lig3 above that conferred by the DNA binding activity of the catalytic domain. The zinc finger of Lig3 may thus be important for rejoining single- or double-strand breaks located within regions of unusual secondary structure, such as multiply damaged sites, highly repetitive sequences, or other intermediates of DNA metabolism.

SSBR Protein Complexes; Pre-Formed or Assembled at Breaks?

There is little information concerning the way in which the protein interactions mediated by XRCC1 are regulated. Each of the interactions described above was identified in cells that had not been exposed to genotoxin, consistent with them occurring at some level constitutively. This does not rule out the possibility that the interactions are triggered by DNA damage, however, since single-strand breaks arise continuously in cells. These possibilities are not mutually exclusive, either. For example, it is possible that XRCC1 interacts to some extent with some or all of its partners independently of DNA damage, but that the abundance of these complexes increases with increased damage. The interaction of XRCC1 with PARP-1 and PARP-2 is certainly stimulated by DNA damage, since it preferentially occurs with the auto-modified enzymes that arise at strand breaks (see above). As discussed above, this is in agreement with a model in which PARP-1/PARP-2 recruit XRCC1 to strand breaks. In contrast, the interaction of most, if not all, cellular Lig3α with XRCC1 appears to be constitutive. This is based on the observed dependence of 80% or more of Lig3α on XRCC1 for cellular stability and on the extent to which the two proteins co-fractionate and co-purify (unpublished observations). It seems unlikely that the majority of cellular Lig3α would bind XRCC1 in response to endogenous levels of DNA damage, and thus it is more likely that they are bound together in a damage-independent manner. It is not yet known how the other interactions mediated by XRCC1 are regulated, but it is clear that only small amounts of the cellular complement of PNK and Polβ cofractionate and co-purify with XRCC1 from protein extracts from "undamaged" cells.

Clearly, further studies are needed to understand how the different interactions mediated by XRCC1 are mediated. However, a currently favoured model is that XRCC1 primarily interacts with its various partners in a sequential manner, at the site of damage.[4,54] This idea underpins the model presented in Figure1, and is based on structural and biochemical studies that suggest that DNA intermediates during BER are passed from one enzyme to the next in a molecular relay driven by sequential protein-protein interactions.[25,26,51] This arrangement, as opposed to the presence of SSBR proteins in large pre-formed complexes, has the advantage of flexibility. For example, the termini of SSBs can differ in their chemistry, depending on their origin, and thereby can differ in the enzymes the require for their processing (Fig. 1c &1d). Thus, by initiating interactions at the site of SSBs, XRCC1 could recruit just those enzyme/s that are necessary. This might also explain why Lig3α is constitutively bound to XRCC1, since a DNA ligase will required at every SSB, irrespective of its origin.

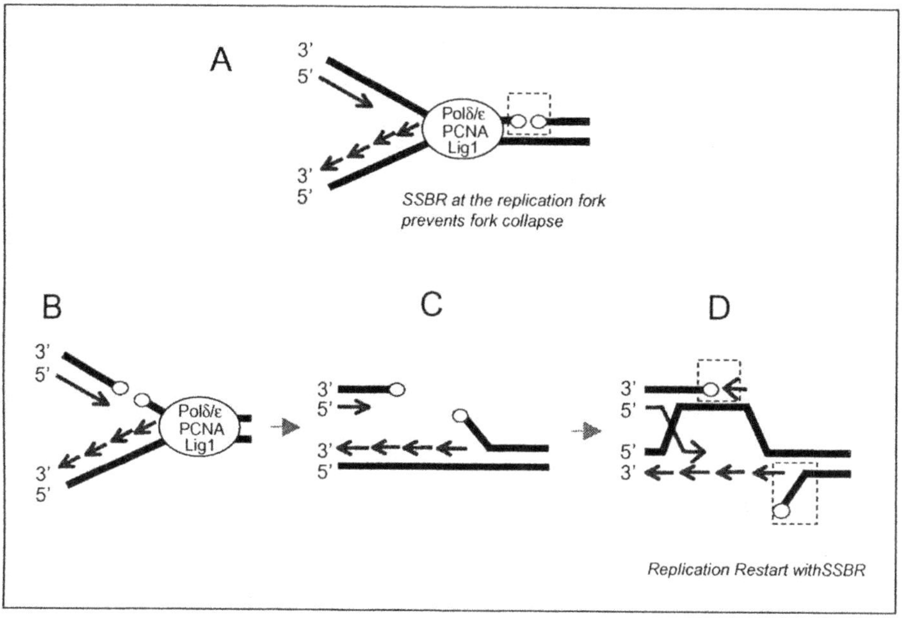

Figure 3. A Model for S/G2-specific XRCC1-dependent SSBR. An unrepaired SSB in the template (black) strand is encountered by the DNA replication machinery (yellow oval) and is repaired ahead of the fork, (A). Alternatively, the SSB collapses the replication fork and creates a DSB, (B ,C). In this case, homologous recombination reactions result in a Holliday junction and restoration of the replication fork, (D). Note that a SSB is still present downstream of the restored replication fork (bottom strand) which, if not repaired, could result in the re-occurrence of a DSB when replication continues. Dotted boxes denote putative sites of SSBR. Red circles denote the possible presence of "damaged", i.e. non-conventional, termini. Please view the color image at www.Eurekah.com.

SSBR Pathways and the Cell Cycle

Site-directed mutagenesis has highlighted the existence of at least three pathways for the repair of SSBs in cells. First, there is a rapid XRCC1-dependent process by which most cellular SSBs induced by alkylating agents[102,106] or H_20_2 are rejoined within 0.5-3-hr, depending on the level of initial damage. This rapid SSBR pathway appears to operate throughout interphase. It seems likely that this process is the cellular equivalent of the repair reactions depicted in Figure 1, which have studied extensively during BER, and to a lesser extent at direct strand breaks, in cell free reactions. For example, the rapid XRCC1-dependent SSBR repair process detected in cells after DNA alkylation is inhibited by disrupting the XRCC1-Lig3α interaction,[101,102] and short-patch BER observed in cell extracts is similarly reduced in cells depleted of Lig3α.[85]

Surprisingly, inhibiting rapid XRCC1-dependent SSBR by preventing the interaction of XRCC1 with Lig3α does not greatly sensitise cells to alkylating agents.[101,102] This appears to be because cells possess a second *XRCC1*-dependent pathway that operates specifically in S/G2. This pathway is a major determinant of cellular resistance to alkylating agents. We have shown that this pathway can remove persisting un-repaired SSBs when the cells enter S/G2, and can thereby compensate for an absence of the global rapid SSBR process. The nature of the *XRCC1*-dependent, S/G2-specific, SSBR process is unknown. However, it seems likely that this process represents the repair of SSBs encountered by a replication fork. We have previously termed this hypothetical process, *replication-coupled SSBR*.[54] Replication-coupled SSBR could be co-ordinated with the replication machinery such that the SSB is removed without a DSB and fork collapse occurring (Fig. 3a). Alternatively, SSBR could occur after the SSB has been converted into a DSB and the fork has collapsed (Fig. 3b-d). In this scenario, since the original SSB that created the DSB is still present after the fork is re-established by homologous recombination, failure to repair the SSB before restarting replication may result in re-occurrence of the DSB and fork collapse. It is not known which proteins function alongside XRCC1 in the S/G2 process. Given that cells employ PCNA, FEN1, and Ligl during chromosomal DNA replication one might speculate that SSBR at the replication fork employs these long-patch repair proteins. The use of Ligl could also explain the observation that the interaction of XRCC1 with Lig3α is dispensable for XRCC1-dependent SSBR during S/G2.[102]

Finally, the rate of SSBR in EM9 cells and EM-C11 cells after exposure to alkylating agents or H_20_2 is reduced, but is not absent during interphase.[41,106] There is thus at least one XRCC1-independent pathway for SSBR. The residual repair of SSBs in XRCC1-mutant cells is relatively slow, and the hypersensitivity of such cells to genotoxins suggests that it cannot fully compensate for the absence of the XRCC1-dependent S/G2 process. Once again, it is not known whether the XRCC1-independent repair process is short or long patch in nature, or how this process relates to the pathways shown in Figure 1. For example, XRCC1-independent repair could simply reflect the inefficient operation of the pathways shown in Figure 1 in the absence of XRCC1, or alternativley it could reflect a completely different process.

Acknowledgements

KWC is funded by an MRC Programme Grant (G0001259/56042).

Reference

1. Lindahl T. Instability and decay of the primary structure of DNA. Nature 1993; 362:709-715.
2. Ward JE. Nature of lesions formed by ionizing radiation. In: Nickoloff JA, Hoekstra MF, eds. DNA Damage and Repair. Vol 2: DNA Repair in Higher Eukaryotes. Totowa: Human Press Inc., 1998:65-84.
3. Nitiss JL. DNA Topoisomerases in DNA repair and DNA damage tolerance. In: Nickoloff JA, Hoekstra MF, eds. DNA Damage and Repair. Totowa: Human Press Inc., 1998:517-537.
4. Whitehouse CJ, Taylor RM, Thistlethwaite A et al. XRCC1 stimulates human polynucleotide kinase activity at damaged DNA termini and accelerates DNA single-strand break repair. Cell 2001; 104:1-11.
5. Thompson LH, West MG. XRCC1 keeps DNA from getting stranded. Mutat.Res 2000; 459:1-18.
6. Tebbs RS, Flannery ML, Meneses JJ et al. Requirement for the Xrcc1 DNA base excision repair gene during early mouse development. Dev Biol 1999; 208):513-529.
7. Dominguez I, Daza P, Natarajan AT et al. A high yield of translocations parallels the high yield of sister chromatid exchanges in the CHO mutant EM9. Mutat Res 1998; 398:67-73.
8. Zhou W, Doetsch PW. Transcription bypass or blockage at single-strand breaks on the DNA template strand: effect of different 3' and 5' flanking groups on the T7 RNA polymerase elongation complex. Biochemistry 1994; 33:14926-14934.
9. Burkle A. Physiology and pathophysiology of poly(ADP-ribosyl)ation. Bioessays 2001;23:795-806.
10. De Murcia G, Menissier DM. Poly(ADP-ribose) polymerase: A molecular nick-sensor. Trends Biochem Sci 1994; 19:172-176.
11. Herceg Z, Wang ZQ. Functions of poly(ADP-ribose) polymerase (PARP) in DNA repair, genomic integrity and cell death. Mutat Res 2001; 477:97-110.
12. Shall S, De Murcia G. Poly(ADP-ribose) polymerase-1: What have we learned from the deficient mouse model? Mutat Res 2000; 460:1-15.
13. Masutani M, Nozaki T, Nakamoto K et al. The response of Parp knockout mice against DNA damaging agents. Mutat Res 2000; 462:159-166.
14. Tulin A, Stewart D, Spradling AC. The Drosophila heterochromatic gene encoding poly(ADP-ribose) polymerase (PARP) is required to modulate chromatin structure during development. Genes Dev 2002; 16:2108-2119.

15. Tulin A, Spradling A. Chromatin loosening by poly(ADP)-ribose polymerase (PARP) at Drosophila puff loci. Science 2003; 299:560-562.
16. Ha HC, Hester LD, Snyder SH. Poly(ADP-ribose) polymerase-1 dependence of stress-induced transcription factors and associated gene expression in glia. Proc Natl Acad Sci USA 2002 ;99:3270-3275.
17. Meisterernst M, Stelzer G, Roeder RG. Poly(ADP-ribose) polymerase enhances activator-dependent transcription in vitro. Proc Natl Acad Sci USA 1997; 94:2261-2265.
18. Simbulan-Rosenthal CM, Ly DH, Rosenthal DS et al. Misregulation of gene expression in primary fibroblasts lacking poly(ADP-ribose) polymerase. Proc Natl Acad Sci USA 2000; 97:11274-11279.
19. Ullrich O, Diestel A, Eyupoglu IY et al. Regulation of microglial expression of integrins by poly(ADP-ribose) polymerase-1. Nat Cell Biol 2001; 3):1035-1042.
20. Vispe S, Yung TM, Ritchot J et al. A cellular defense pathway regulating transcription through poly(ADP-ribosyl)ation in response to DNA damage. Proc Natl Acad Sci USA 2000; 97:9886-9891.
21. Chiarugi A, Moskowitz MA. Cell biology. PARP-1—A perpetrator of apoptotic cell death? Science 2002; 297:200-201.
22. Yu SW, Wang H, Poitras MF et al. Mediation of poly(ADP-ribose) polymerase-1-dependent cell death by apoptosis-inducing factor. Science 2002; 297:259-263.
23. Schreiber V, Ame JC, Dolle P et al. Poly(ADP-ribose) polymerase-2 (PARP-2) is required for efficient base excision DNA repair in association with PARP-1 and XRCC1. J Biol Chem 2002; 277:23028-23036.
24. Le Cam E, Fack F, Menissier-de Murcia J et al. Conformational analysis of a 139 base-pair DNA fragment containing a single-stranded break and its interaction with human poly(ADP-ribose) polymerase. J Mol Biol 1994; 235:1062-1071.
25. Mol CD, Izumi T, Mitra S et al. DNA-bound structures and mutants reveal abasic DNA binding by APE1 and DNA repair coordination [corrected]. Nature 2000; 403:451-456.
26. Wilson SH, Kunkel TA. Passing the baton in base excision repair. Nat Struct Biol 2000; 7:176-178.
27. Ferro AM, Olivera BM. Poly(ADP-ribosylation) in vitro. Reaction parameters and enzyme mechanism. J Biol Chem 1982; 257:7808-7813.
28. Zahradka P, Ebisuzaki K. A shuttle mechanism for DNA-protein interactions. The regulation of poly(ADP-ribose) polymerase. Eur J Biochem 1982; 127:579-585.
29. Lindahl T, Satoh MS, Poirier GG, Klungland A. Post-translational modification of poly(ADP-ribose) polymerase induced by DNA strand breaks. Trends Biochem Sci 1995; 20:405-411.
30. Sanderson RJ, Lindahl T. Down-regulation of DNA repair synthesis at DNA single-strand interruptions in poly(ADP-ribose) polymerase-1 deficient murine cell extracts. DNA Repair (Amst) 2002; 1:547-558.
31. Satoh MS, Lindahl T. Role of poly(ADP-ribose) formation in DNA repair. Nature 1992; 356:356-358.
32. Vodenicharov MD, Sallmann FR, Satoh MS et al. Base excision repair is efficient in cells lacking poly(ADP-ribose) polymerase 1. Nucleic Acids Res 2000; 28:3887-3896.
33. Masson M, Niedergang C, Schreiber V et al. XRCC1 is specifically associated with poly(ADP-ribose) polymerase and negatively regulates its activity following DNA damage. Mol Cell Biol 1998; 18:3563-3571.
34. Caldecott KW, Aoufouchi S, Johnson P et al. XRCC1 polypeptide interacts with DNA polymerase beta and possibly poly (ADP-ribose) polymerase, and DNA ligase III is a novel molecular 'nick-sensor' in vitro. Nucleic Acids Res 1996; 24:4387-4394.
35. Okano S, Kanno S, Nakajima S et al. Cellular responses and repair of single-strand breaks introduced by UV damage endonuclease in mammalian cells. J Biol Chem 2000; 275:32635-32641.
36. Dantzer F, de La RG, Menissier-de Murcia J et al. Base excision repair is impaired in mammalian cells lacking Poly(ADP-ribose) polymerase-1. Biochemistry 2000; 39:7559-7569.
37. Cantoni O, Murray D, Meyn RE. Induction and repair of dna single-strand breaks in em9 mutant cho cells treated with hydrogen-peroxide. Chemico-Biological Interactions 1987; 63:29-38.
38. Christie NT, Cantoni O, Evans RM et al. Use of mammalian DNA repair-deficient mutants to assess the effects of toxic metal compounds on DNA. Biochem Pharmacol 1984; 33:1661-1670.
39. Churchill ME, Peak JG, Peak MJ. Correlation between cell-survival and dna single-strand break repair proficiency in the chinese-hamster ovary cell-lines aa8 and em9 irradiated with 365-nm ultraviolet-a radiation. Photochemistry and Photobiology 1991; 53:229-236.

40. Churchill ME, Peak JG, Peak MJ. Repair of near-visible-light-induced and blue-light-induced dna single-strand breaks by the cho cell-lines aa8 and em9. Photochemistry and Photobiology 1991; 54:639-644.
41. Thompson LH, Brookman KW, Dillehay LE et al. A CHO-cell strain having hypersensitivity to mutagens, a defect in DNA strand-break repair, and an extraordinary baseline frequency of sister-chromatid exchange. Mutat Res 1982; 95:427-440.
42. Dillehay LE, Thompson LH, Carrano AV. DNA-strand breaks associated with halogenated pyrimidine incorporation. Mutat Res 1984; 131:129-136.
43. Pleschke JM, Kleczkowska HE, Strohm M et al. Poly(ADP-ribose) binds to specific domains in DNA damage checkpoint proteins. J Biol Chem 2000; 275:40974-40980.
44. Satoh MS, Poirier GG, Lindahl T. NAD(+)-dependent repair of damaged DNA by human cell extracts. J Biol Chem 1993; 268:5480-5487.
45. Satoh MS, Poirier GG, Lindahl T. Dual function for poly(ADP-ribose) synthesis in response to DNA strand breakage. Biochemistry 1994; 33:7099-7106.
46. Bennett RA, Wilson DM, III, Wong D et al. Interaction of human apurinic endonuclease and DNA polymerase beta in the base excision repair pathway. Proc Natl Acad Sci USA 1997; 94:7166-7169.
47. Caldecott KW. DNA single-strand break repair and spinocerebellar ataxia. Cell 2003; 112:7-10.
48. Trucco C, Oliver FJ, De Murcia G et al. DNA repair defect in poly(ADP-ribose) polymerase-deficient cell lines. Nucleic Acids Res 1998; 26:2644-2649.
49. Le Page F, Schreiber V, Dherin C et al. Poly(ADP-ribose) polymerase-1(PARP-1) is required in murine cell lines for base excision repair of oxidative DNA damage in absence of DNA polymerase beta. J Biol Chem 2003.
50. Prasad R, Lavrik OI, Kim SJ et al. DNA polymerase beta-mediated long patch base excision repair. Poly(ADP-ribose)polymerase-1 stimulates strand displacement DNA synthesis. J Biol Chem 2001; 276:32411-32414.
51. Mol CD, Hosfield DJ, Tainer JA. Abasic site recognition by two apurinic/apyrimidinic endonuclease families in DNA base excision repair: the 3' ends justify the means. Mutat Res 2000; 460:211-229.
52. Nakamura J, La DK, Swenberg JA. 5'-nicked apurinic/apyrimidinic sites are resistant to beta-elimination by beta-polymerase and are persistent in human cultured cells after oxidative stress. J Biol Chem 2000; 275:5323-5328.
53. Xu YJ, Kim EY, Demple B. Excision of C-4'-oxidized deoxyribose lesions from double-stranded DNA by human apurinic/apyrimidinic endonuclease (Ape1 protein) and DNA polymerase beta. J Biol Chem 1998; 273:28837-28844.
54. Caldecott KW. Mammalian DNA single-strand break repair: An X-ra(y)ted affair. Bioessays 2001; 23:447-555.
55. Chen DS, Herman T, Demple B. Two distinct human DNA diesterases that hydrolyze 3'-blocking deoxyribose fragments from oxidized DNA. Nucleic Acids Res 1991; 19:5907-5914.
56. Izumi T, Hazra TK, Boldogh I et al. Requirement for human AP endonuclease 1 for repair of 3'-blocking damage at DNA single-strand breaks induced by reactive oxygen species. Carcinogenesis 2000; 21:1329-1334.
57. Winters TA, Henner WD, Russell PS et al. Removal of 3'-phosphoglycolate from DNA strand-break damage in an oligonucleotide substrate by recombinant human apurinic/apyrimidinic endonuclease 1. Nucleic Acids Res 1994; 22:1866-1873.
58. Demple B, Harrison L. Repair of oxidative damage to DNA: Enzymology and biology. Annu Rev Biochem 1994; 63:915-948.
59. Dianova II, Bohr VA, Dianov GL. Interaction of human AP endonuclease 1 with flap endonuclease 1 and proliferating cell nuclear antigen involved in long-patch base excision repair. Biochemistry 2001; 40:12639-12644.
60. Ranalli TA, Tom S, Bambara RA. AP endonuclease 1 coordinates flap endonuclease 1 and DNA ligase I activity in long patch base excision repair. J Biol Chem 2002; 277:41715-41724.
61. Tom S, Ranalli TA, Podust VN et al. Regulatory roles of p21 and apurinic/apyrimidinic endonuclease 1 in base excision repair. J Biol Chem 2001; 276:48781-48789.

62. Vidal AE, Boiteux S, Hickson ID et al. XRCC1 coordinates the initial and late stages of DNA abasic site repair through protein-protein interactions. EMBO J 2001; 20:6530-6539.
63. Chou KM, Cheng YC. An exonucleolytic activity of human apurinic/apyrimidinic endonuclease on 3' mispaired DNA. Nature 2002; 415:655-659.
64. Jilani A, Ramotar D. Purification and partial characterization of a DNA 3'-phosphatase from Schizosaccharomyces pombe. Biochemistry 2002; 41:7688-7694.
65. Karimi-Busheri F, Lee J, Tomkinson AE et al. Repair of DNA strand gaps and nicks containing 3'-phosphate and 5'-hydroxyl termini by purified mammalian enzymes. Nucleic Acids Res 1998; 26:4395-4400.
66. Karimi-Busheri F, Weinfeld M. Purification and substrate specificity of polydeoxyribonucleotide kinases isolated from calf thymus and rat liver. J.Cell Biochem. 1997; 64:258-272.
67. Inamdar KV, Pouliot JJ, Zhou T et al. Conversion of phosphoglycolate to phosphate termini on 3' overhangs of DNA double strand breaks by the human tyrosyl-DNA phosphodiesterase hTdp1. J Biol Chem 2002; 277:27162-27168.
68. Meijer M, Karimi-Busheri F, Huang TY et al. Pnk1, a DNA kinase/phosphatase required for normal response to DNA damage by gamma-radiation or camptothecin in Schizosaccharomyces pombe. J Biol Chem 2002; 277:4050-4055.
69. Beard WA, Wilson SH. Structural design of a eukaryotic DNA repair polymerase: DNA polymerase beta. Mutat Res 2000; 460:231-244.
70. Wilson SH. Mammalian base excision repair and DNA polymerase beta. Mutat Res 1998; 407:203-215.
71. DiGiuseppe JA, Dresler SL. Bleomycin-induced DNA repair synthesis in permeable human fibroblasts: mediation of long-patch and short-patch repair by distinct DNA polymerases. Biochemistry 1989; 28:9515-9520.
72. Miller MR, Chinault DN. The roles of DNA polymerases alpha, beta, and gamma in DNA repair synthesis induced in hamster and human cells by different DNA damaging agents. J Biol Chem 1982; 257:10204-10209.
73. Miller MR, Chinault DN. Evidence that DNA polymerases alpha and beta participate differentially in DNA repair synthesis induced by different agents. J Biol Chem 1982; 257:46-49.
74. Winters TA, Russell PS, Kohli M et al. Determination of human DNA polymerase utilization for the repair of a model ionizing radiation-induced DNA strand break lesion in a defined vector substrate. Nucleic Acids Res 1999; 27:2423-2433.
75. Fortini P, Pascucci B, Parlanti E et al. Different DNA polymerases are involved in the short- and long-patch base excision repair in mammalian cells. Biochemistry 1998;37:3575-3580.
76. Frosina G, Fortini P, Rossi O et al. Repair of abasic sites by mammalian cell extracts. Biochem J 1994; 304(Pt 3):699-705.
77. Frosina G, Fortini P, Rossi O et al. Two pathways for base excision repair in mammalian cells. J Biol Chem 1996; 271:9573-9578.
78. Matsumoto Y, Kim K, Hurwitz J et al. Reconstitution of proliferating cell nuclear antigen-dependent repair of apurinic/apyrimidinic sites with purified human proteins. J Biol Chem 1999; 274:33703-33708.
79. Pascucci B, Stucki M, Jonsson ZO et al. Long patch base excision repair with purified human proteins. DNA ligase I as patch size mediator for DNA polymerases delta and epsilon. J Biol Chem 1999; 274:33696-33702.
80. Matsumoto Y, Kim K. Excision of deoxyribose phosphate residues by DNA polymerase beta during DNA repair. Science 1995; 269:699-702.
81. Klungland A, Lindahl T. Second pathway for completion of human DNA base excision-repair: reconstitution with purified proteins and requirement for DNase IV (FEN1). EMBO J 1997; 16:3341-3348.
82. Matsumoto Y, Kim K, Bogenhagen DF. Proliferating cell nuclear antigen-dependent abasic site repair in Xenopus laevis oocytes: an alternative pathway of base excision DNA repair. Mol Cell Biol 1994; 14:6187-6197.
83. Prasad R, Dianov GL, Bohr VA et al. FEN1 stimulation of DNA polymerase beta mediates an excision step in mammalian long patch base excision repair. J Biol Chem 2000; 275:4460-4466.

84. Gary R, Kim K, Cornelius HL et al. Proliferating cell nuclear antigen facilitates excision in long-patch base excision repair. J Biol Chem 1999; 274:4354-4363.
85. Cappelli E, Taylor R, Cevasco M et al. Involvement of XRCC1 and DNA ligase III gene products in DNA base excision repair. J Biol Chem 1997; 272:23970-23975.
86. Levin DS, McKenna AE, Motycka TA et al. Interaction between PCNA and DNA ligase I is critical for joining of Okazaki fragments and long-patch base-excision repair. Curr Biol 2000; 10:919-922.
87. Kubota Y, Nash RA, Klungland A et al. Reconstitution of DNA base excision-repair with purified human proteins: Interaction between DNA polymerase beta and the XRCC1 protein. EMBO J 1996; 15:6662-6670.
88. Marintchev A, Mullen MA, Maciejewski MW et al. Solution structure of the single-strand break repair protein XRCC1 N-terminal domain. Nat Struct Biol 1999; 6:884-893.
89. Marintchev A, Robertson A, Dimitriadis EK et al. Domain specific interaction in the XRCC1-DNA polymerase beta complex. Nucleic Acids Res 2000; 28:2049-2059.
90. Marintchev A, Gryk MR, Mullen GP. Site-directed mutagenesis analysis of the structural interaction of the single-strand-break repair protein, X-ray cross-complementing group 1, with DNA polymerase beta. Nucleic Acids Res 2003; 31:580-588.
91. Rice PA. Holding damaged DNA together. Nat Struct Biol 1999; 6:805-806.
92. Caldecott KW, Mckeown CK, Tucker JD et al. An interaction between the mammalian DNA repair protein XRCC1 and DNA ligase III. Mol Cell Biol 1994; 14:68-76.
93. Caldecott KW, Tucker JD, Stanker LH et al. Characterization of the XRCC1-DNA ligase III complex in vitro and its absence from mutant hamster cells. Nucleic Acids Res 1995; 23:4836-4843.
94. Chen J, Tomkinson AE, Ramos W et al. Mammalian DNA ligase III: Molecular cloning, chromosomal localization, and expression in spermatocytes undergoing meiotic recombination. Mol Cell Biol 1995; 15:5412-5422.
95. Mackey ZB, Ramos W, Levin DS et al. An alternative splicing event which occurs in mouse pachytene spermatocytes generates a form of DNA ligase III with distinct biochemical properties that may function in meiotic recombination. Mol Cell Biol 1997; 17:989-998.
96. Wei YF, Robins P, Carter K et al. Molecular cloning and expression of human cDNAs encoding a novel DNA ligase IV and DNA ligase III, an enzyme active in DNA repair and recombination. Mol Cell Biol 1995; 15:3206-3216.
97. Nash RA, Caldecott KW, Barnes DE et al. XRCC1 protein interacts with one of two distinct forms of DNA ligase III. Biochemistry 1997; 36:5207-5211.
98. Zhang XD, Morera S, Bates PA et al. Structure of an XRCC1 BRCT domain: a new protein-protein interaction module. EMBO J 1998; 17:6404-6411.
99. Shen NR, Zdzienicka MZ, Mohrenweiser H et al. Mutations in hamster single-strand break repair gene XRCC1 causing defective DNA repair. Nucl Acids Res 1998; 26:1032-1037.
100. Ljungquist S, Kenne K, Olsson L et al. Altered dna-ligase-iii activity in the cho em9 mutant. Mut Res 1994; 314:177-186.
101. Moore DJ, Taylor RM, Clements P et al. Mutation of a BRCT domain selectively disrupts DNA single-strand break repair in noncycling Chinese hamster ovary cells. Proc Natl Acad Sci USA 2000; 97:13649-13654.
102. Taylor RM, Moore DJ, Whitehouse J et al. A cell cycle-specific requirement for the XRCC1 BRCT II domain during mammalian DNA strand break repair. Mol Cell Biol 2000; 20:735-740.
103. Taylor RM, Whitehouse J, Cappelli E et al. Role of the DNA ligase III zinc finger in polynucleotide binding and ligation. Nucleic Acids Res 1998; 26:4804-4810.
104. Taylor RM, Whitehouse CJ, Caldecott KW. The DNA ligase III zinc finger stimulates binding to DNA secondary structure and promotes end joining [In Process Citation]. Nucleic Acids Res 2000; 28:3558-3563.
105. Mackey ZB, Niedergang C, Meenissier de Murcia J et al. DNA ligase III is recruited to DNA strand breaks by a zinc finger motif homologous to that of poly(ADP-ribose) polymerase—Identification of two functionally distinct DNA binding regions within DNA ligase III. J Biolog Chem 1999; 274:21679-21687.
106. Taylor RM, Thistlethwaite A, Caldecott KW. Central role for the XRCC1 BRCT I domain in mammalian DNA single-strand break repair. Mol Cell Biol 2002; 22:2556-2563.

Chapter 4

The Detection and Repair of Nucleotide Damage and Coupling to Major Cellular Regulatory Processes in Mammalian Cells

James E. Cleaver

Abstract

DNA repair serves to restore two of the main macromolecular functions in the cell: DNA replication and transcription. If either are irrevocably blocked cell death in the form of apoptosis or permanent arrest ensues. Damage in the form of large modifications, pyrimidine dimers from UV light or DNA adducts from carcinogenic chemicals, are repaired by nucleotide excision repair (NER). This replaces the damage with a 27-29 nucleotide patch by a combination of damage recognition, verification, DNA unwinding, 3' and 5' cleavage and DNA replication. Two main pathways have been formally identified: global repair in which initial damage recognition occurs through dedicated proteins XPC/HR23B and XPE (DDB1/DDB2); transcription coupled repair (TCR) in which damage recognition occurs as a result of RNA pol II arrest. Many transcription factors are involved in TCR, and two proteins dedicated to TCR, CSA and CSB, are required for RNA pol II ubiquitination and are mutated in the sun-sensitive neurodegenerative disease Cockayne syndrome. Overlap of NER with repair of oxidative damage may be responsible for the neural degeneration in CS. Transcription of the global repair components XPC/HR23B and XPE (DDB1/DDB2) are induced by p53 transactivation, but transcription of the majority of the NER components are not induced by damage. Both branches of NER interact with the proteasome degradation pathway, although this may involve monoubiquitination for modification of protein function as well as protein degradation. Unrepaired damage is replicated by damage-specific polymerases, mainly Pol H, I, and K of the class Y low fidelity polymerases, with additional contribution by high fidelity polymerases Pol Z and terminal dCMP transferase hRev1, and other replication and recombination factors. NER is vitally important for protection against solar-induced damage that can lead to immunosuppression and skin cancer, including melanoma, and has important implications for other human and mammalian cancers.

Introduction

DNA repair was investigated during the initial decades of discovery as a set of self-contained systems for editing and correcting DNA damage that operated in their own microcosms, impervious to most outside cellular and tissue influences.[1] The recent explosive growth of DNA repair studies has overturned this approach, and instead DNA repair is now viewed as a central

Eukaryotic DNA Damage Surveillance and Repair, edited by Keith W. Caldecott.

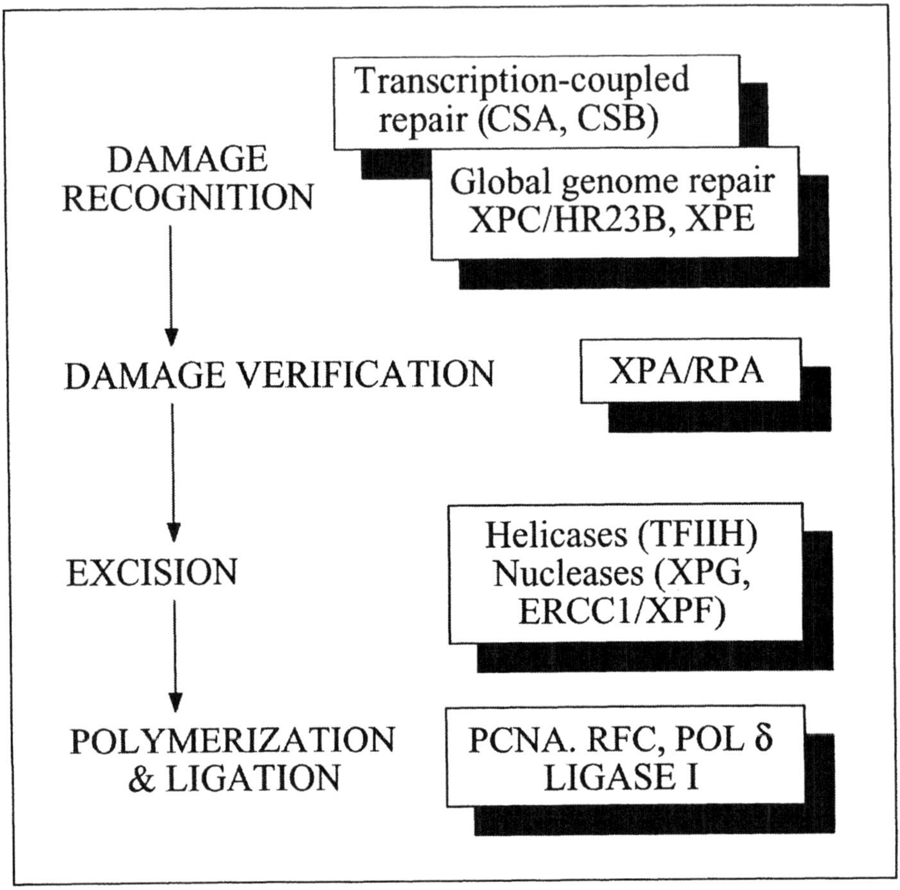

Figure 1. The sequence of nucleotide excision repair from damage recognition, verification, excision, polymerization and ligation. The molecular components of each step are shown in boxes to the right of the pathway.

feature of cellular homeostasis and the different repair systems are dynamically intertwined with one another and with normal cell metabolism. Repair genes make up a significant part of every sequenced genome and the varied processes interact with most major cellular transactions of cell cycle regulation and checkpoint control, mitosis, transcription, DNA replication, telomerase maintenance and proteasome-mediated protein degradation.[2] This review will discuss repair of large DNA adducts, especially those produced by UV light. The response of cells to large DNA adducts primarily involves nucleotide excision repair (NER) but recruits or is linked to components of base excision repair (BER), double strand break repair (DSBR), mismatch repair (MMR), DNA replication and recombination. Most of the main components of the NER system are now known: the genes are cloned, chromosomal locations identified, proteins available in recombinant form, functions identified, and mutations detected in the human diseases xeroderma pigmentosum (XP), cockayne syndrome (CS), trichothiodystrophy (TTD). These have been summarized in the web-site www.xpmutations.org and in (Table1). Detailed description of the DNA repair deficient diseases and their pathology have been described elsewhere.[3-6]

Table 1. The main genes and functions of the NER system

Gene	Chrom Location[1]	Protein	Function	Partner(s)
XPA	9q34.1	273aa	DNA damage binding	RPA
XPB	2q21	782aa	3′-5′ helicase	TFIIH
XPC	3p25.1	940aa	DNA damage binding	hHR23B
XPD	19q13.2	760aa	5′-3′ helicase	TFIIH
XPE	11p11-12	427aa	DNA damage binding (DDB2)	DBB1(p127)
XPF	16p13.3	916aa	Endonuclease 5′to damage	ERCC1
XPG	13q32-33	1186aa	Endonuclease 3′to damage	Thymine glycosylase (nth)
XPV	6p21	713aa	Low fidelity DNA polymerase (pol H or η)	PCNA
CSA	5q11.2	396aa	WD protein, ubiquitination of RNA pol II	
CSB	10q23	1493aa	DNA dependent ATPase, ubiqu itination of RNA pol II	

For further details on chromosome locations and sequence see www.ensembl.org and for mutations in the genes see www.xpmutations.org.

An early assertion described an intimate connection between DNA replication and the capacity of cells to carry out NER.[7] This assertion was soon eclipsed by the demonstration of a strong connection between NER and gene transcription.[8] Transcribed genes and the transcribed strands were shown to be more rapidly repaired,[9] and the gene products of Cockayne syndrome (CS) were shown to be dedicated to this enhanced repair. As a result a specific branch of NER was named "Transcription coupled repair (TCR)" that is distinct from repair of the remainder of the genome, called "Global genome repair" (GGR) (Fig. 1). Major components of NER were also identified as components of the basal transcription factor TFIIH thereby fully integrating gene transcription and DNA repair.[10]

The idea of linkage between replication and repair has resumed importance, recently. A large protein complex (BASC) can be identified in mammalian cell nuclei that contains a close association of many components of DNA replication, repair and recombination.[11] Not only is there interaction between these proteins, but many of the interactions occur within macroscopic, visible protein complexes that may be dynamic entities that come together to increase their specificity and efficiency.[11] In addition, NER defects and replication defects in the human disease XP result in indistinguishable clinical symptoms of elevated skin cancer risks.[12] Older evidence suggested that recovery of DNA replication after UV damage was correlated with the capacity for NER,[13] and required the activity of the Cockayne syndrome genes, *CSA* and *B*, more commonly associated with TCR.[14] Caffeine sensitivity of UV damaged cells is expressed specifically in the S phase and is associated with enhanced replication arrest,[15,16] but this sensitization is also correlated with NER capacity.[17] MMR is primarily involved in correcting DNA replication and deamination errors [18-20] but is also a determining factor in strand choice during TCR in bacteria and some cancer cells[21-23] and in mediating UV damage, repair and apoptosis.[24]

There is therefore ample evidence for common elements involved in many DNA transactions associated with DNA damage responses and normal cellular processes of macromolecular synthesis and degradation. At its simplest, two functions are essential to cells under conditions of stress: one is the ability to restore transcription ability; the second is to restore DNA replication and prevent degradation of arrested replication forks. If either of these become insurmountable then the outcome is apoptosis[25] or persistent replication arrest.[26] Although in prokaryotes many DNA repair enzymes are inducible in response to damage,[1] in eukaryotes a different strategy appears to predominate. Most of the eukaryotic enzymes required for repair are constitutive,[27,28] except for a small number in human cells that are p53-responsive such as *DDB2* and *GADD45*.[29] The genes that are induced by damage appear to be more generalized stress response genes and not DNA repair genes responsible for resistance to the inducing agents.[27,28,30] In human keratinocytes, for example, the gene families induced by UVB represent inflammatory and stress response genes, and those involved in the mechanism of gene transcription and proteasome-mediated degradation.[30] Modification of enzyme activity by phosphorylation, however, represents a major strategy by which eukaryotic cells regulate their response to cellular and DNA damage.[31-35] These controls serve to manage the linkage of cell cycle checkpoints with DNA repair, and the sequential handing on of repair intermediates from one set of enzymes to another until the whole process is complete (Fig. 1).[36]

This is now a suitable occasion to review what is currently known about NER and its involvement with transcription, DNA replication and other major cellular regulatory processes, with special emphasis on more recent information. We will discuss the core NER process, its two major branches of GGR and TCR, and secondary levels of regulation by phosphorylation and proteolysis.

Types of Damage Repaired by NER

The NER system recognizes and repairs DNA damage that consists of ultraviolet (UV)-induced photoproducts and large DNA adducts. These include the cyclobutane [5-5], [6-6] pyrimidine dimers and the [6-4] pyrimidine pyrimidinone dimers that can involve T and C pyrimidines. The [6-4] photoproduct can further photoisomerase to the Dewar photoproduct at longer UV wavelengths, and cytosine in dimers has an increased propensity to deaminate contributing to C to T mutagenesis.[37] Chemical adducts include those produced by N-acetoxy-N-acetyl aminofluorene (AAAF), benzo(a)pyrene, aflatoxin, photoactivated psoralens, and cis-platinum. An oxidative purine product, 5',8-purine cyclodeoxynucleoside, that may accumulate in neurological tissue also requires the NER system for repair.[38] The NER system can even recognize DNA triplexes formed by the binding of short oligonucleotides to double stranded DNA.[39]

In most prokaryotes and eukaryotes, the UV photoproducts can also be directly reversed by photolyase enzymes, which cleave the [5-5], [6-6] bonds of the cyclobutane dimer, or the [6-4] linkage of the pyrimidine pyrimidinone dimer.[40] These enzymes channel absorbed blue or visible light quanta into electron transport though several photoreceptor cofactors, to cleave the photoproduct linkages.[41] In mammalian cells, however, monomerization of the photoproducts does not occur,[40,42,43] and the corresponding genes have evolved into blue light receptors involved in photoperiodism.[44]

Most X-ray damage is processed differently using single strand and double strand break repair pathways.[45-51] Most single base damage caused by alkylating agents or oxidative reactions are repaired by short or long patch base excision repair (BER) and alkyltransferase.[1,47,52] Akylations in single stranded DNA can also be repaired by oxidative demethylation and release from the damaged base by the gene product of *AlkB*, that is widespread in prokaryotes and eukaryotes, including human cells.[53-55] The distinctions between the various repair systems are not absolute, however, and there are overlaps in the substrate specificity of these various repair systems and variations in the efficiencies and sites of action on DNA in differing metabolic states.

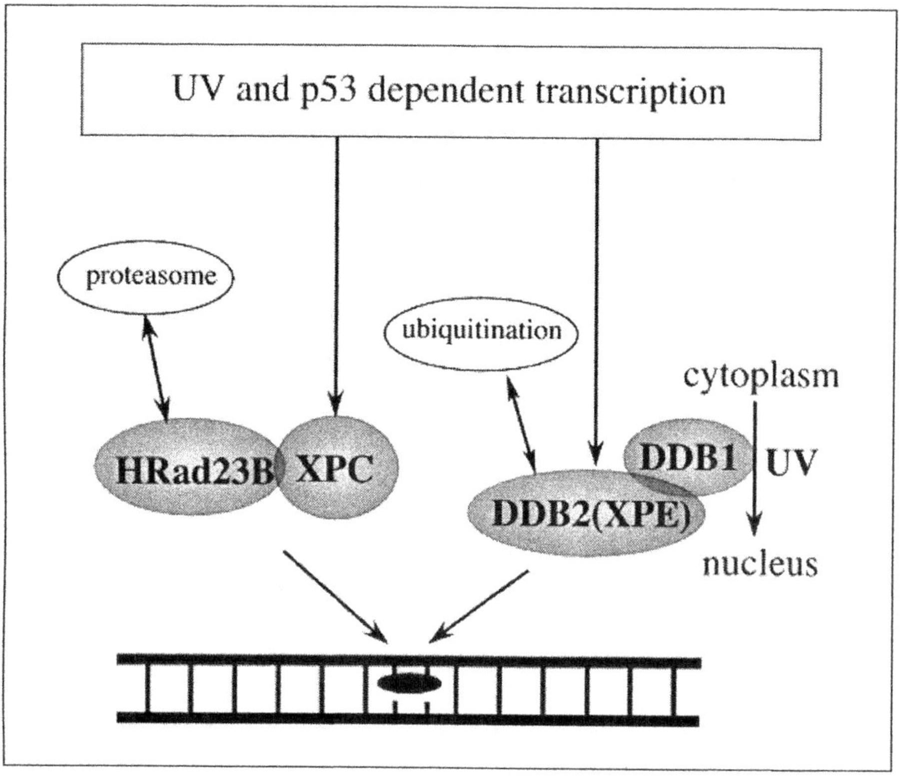

Figure 2. Damage recognition by the global genome repair factors. The primary recognition protein is the heterodimer XPC/HR23B, but additional role is played by the DDB1/2 heterodimer. Transcription of each of these components is stimulated by damage and p53. They also interact with the ubiquitination and proteasome pathways. DDB1 normally resides in the cytoplasm and migrates to the nucleus after UV damage.

Recognition of Damage in Nontranscribing Regions of the Genome

The initial damage recognition factors uniquely required for GGR are the XPC and XPE DNA binding proteins (Fig. 2). XPC is found in association with a cofactor HR23B, one of the two human homologs of the yeast Rad23 gene product. The XPC/HR23B complex [56,57] is the earliest damage detector to initiate NER in nontranscribed DNA, acting to stabilize subsequent XPA/RPA binding to the damaged site, with a high affinity for the [6-4]PP.[58,59] The XPC protein may be required for transient nucleosome unfolding during NER.[60] In XPC cells that lack this complex, repair is confined to small transcriptionally active islands of the genome, where repair occurs at normal rates, surrounded by oceans of unrepaired transcriptionally silent DNA.[61-63] These regions have been dramatically visualized by immunofluorescence.[64] The XPC/HR23B complex is also associated with CEN2, a protein that stabilizes XPC and which is also involved in centrosome duplication, suggesting a direct link between GGR and cell division.[65] There is evidence that the constitutive level of expression of XPC is controlled by p53, and the protein can be induced by UV irradiation.[30] Although XPC/HR23B is an important damage recognition protein and acts to open up a damaged region, if the damage already presents itself with a significant degree of unpaired bases the requirement for XPC/HR23B is relaxed.[66-68]

Another factor that plays a role in GGR is the DNA damage binding protein (DDB) that was one of the earliest proteins to be identified with damage binding properties, and is linked to the XP group E complementation group. The XPE group includes patients who are mildly to moderately affected and whose cells carry out near normal levels of NER.[69] The XPE gene product involves a dimeric protein having subunits of 127 kDa (DDB1) and 48 kDa (DDB2).[70-73] Mutations that are diagnostic for XPE are located in the DDB2 subunit (see http://www.xpmutations.org for a summary of XP and CS mutations). Early studies presented a confused picture in which cells originally assigned to the XPE complementation group could be either DDB^+ or DDB^-. More recent work has clarified the situation because cells without mutations in DDB2 actually belong to other groups, after further analysis.[69,74-76]

The DDB1/2 heterodimer is involved in the recognition of damaged DNA and has the capacity to bind to UV-damaged DNA with a chemically defined footprint.[77] The binding is detectable by an electrophoretic mobility shift assay (EMSA) when DDB is bound to a defined oligonucleotide containing a damaged site.[70] But a precise biochemical role for DDB1/2 in the pathway of excision repair has proved elusive.[5] Reconstitution of excision repair using purified proteins showed no requirement for DDB1/2 for damage removal and repair synthesis from synthetic substrates lacking chromatin structure.[78-80] Excision repair in XPE cell extracts can be stimulated by single-strand binding protein RPA, but this stimulation is not specific for XPE.[81] Moreover, no mutations have been found in the subunits of RPA.[78,81,82] These results suggest that DDB1/2 may not be directly involved in excision repair directly, although it is involved in UV-induced mutagenesis.[83]

DDB1 (p127) is present in excess over its partner, and is predominantly cytoplasmic but translocates to the nucleus after UV irradiation.[84] Mutations in XPE that are in DDB2, also prevent the accumulation of DDB1 in the nucleus.[85] DDB2 expression is induced by UV light through p53 transactivation in human, but not mouse or hamster, cells.[72,86,87] The consensus DNA-binding site for p53 is present in the human p48 promoter region but has either diverged or is absent from the mouse homolog.[29] These observations provide a partial explanation for early observations that excision repair can be low in some mouse cell strains.[88-90]

DDB1/2 protein also functions as a transcriptional activator through E2F1.[85,91] The cytoplasmic DDB1 protein also interacts with a range of viral proteins[92,93] and the amyloid precursor protein involved with Alzheimers.[94]

Recognition of Damage in Transcriptionally Active Regions of the Genome

Pyrimidine dimers are excised more rapidly from actively transcribed genes, especially from the DNA strand used as the template for transcription.[8,9,21,95,96] A detailed base by base analysis of excision in the promoter and early region of the *PGK1* gene, using ligation-mediated PCR, revealed that excision is slow where binding proteins interact with the promoter but increases immediately after the ATG start site for transcription.[97] The initial damage recognition mechanism for TCR may be the stalled RNA pol II, itself, although a potentially large number of gene products play a role in mediating the increased rate of repair in transcriptionally active genes (Fig. 3).[8,9] At present the process of TCR has not been reassembled in vitro, so the exact components and their respective roles are still unclear (Fig. 3). The breast cancer susceptibility gene *Brca1*, for example, is required for TCR of oxidative damage to DNA, though whether it is required for TCR of UV damage is unclear.[98,99]

The strand preference for repair in some cell types appears to involve the MMR system, whose major role is correcting mis-matched bases generated through replication errors or deamination of cytosine,[19,20] and can discriminate parental from newly synthesized DNA strands.[21,22] Early evidence indicated that TCR was reduced in MMR defective *E. coli* and human cancer

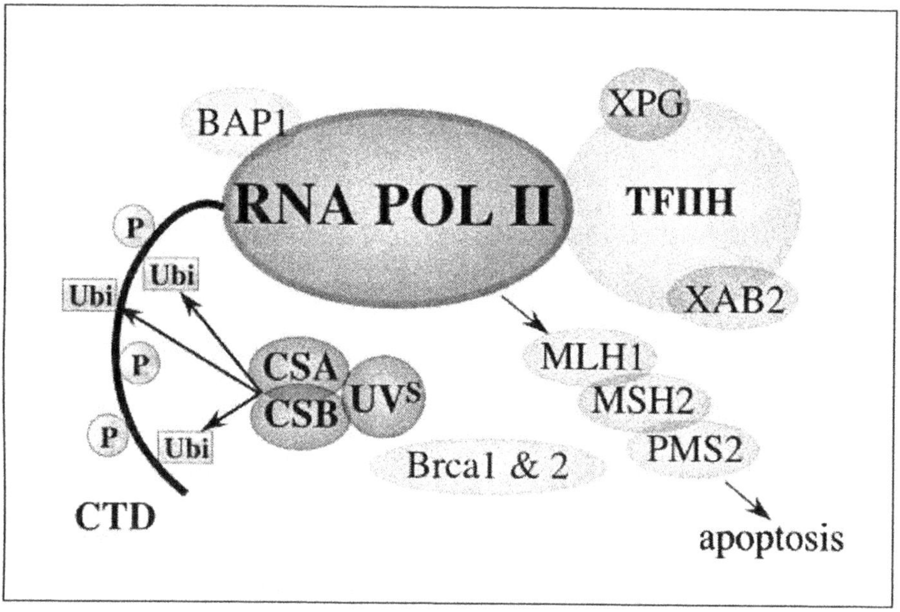

Figure 3. Known participants in transcription coupled damage recognition. These components have all been shown to affect rates of TCR, although their precise structural and functional roles are unclear at present. Mismatch repair components (MLH1, MSH2, PMS2) play an important role in the increased rate of repair on the transcribed strand. The CSA and B proteins, and the uncloned UV^s protein, are involved in ubiquitination of the C-terminal domain (CTD) of RNA pol II, which is required for increased repair of the transcribed strand and for recovery of transcription after UV damage. The XPB and D components of TFIIH are also important. Ubi, P: ubiquitin and phosphate moieties on the CTD.

cells.[21-23,95,96] Subsequent studies in yeast[100] and in mouse embryo fibroblasts from Msh2-defective mice[101] did not demonstrate a role for MMR in TCR. Subsequent research in NER defective mice demonstrated that during carcinogenesis there is selection for MMR defective cells with concomitant reduction in apoptosis.[102] These results are consistent if the main function of MMR in UVB responses is to couple RNA pol II arrest with apoptosis.[24,25] During UVB carcinogenesis, selection in favor of reduced apoptosis brings with it reduction in one or more steps of the apoptotic signaling pathway, which would include MMR.

The arrest of RNA pol II at sites of DNA damage during transcription elongation plays a central role for damage recognition in TCR.[103] The blocked RNA pol II masks damaged sites and must be removed, or backed off, to alleviate the arrested transcription and provide access for repair enzymes. Persistently blocked RNA pol II can be a signal for UV-induced apoptosis.[25] The processes brought into play at the arrested RNA pol II enhance the overall accessibility of the damaged site and result in more rapid repair than in nontranscribed regions of the genome. Transcription arrest is associated with hyperphosphorylation of the C-terminal domain (CTD) of RNA pol II and subsequent ubiquitination of a small fraction of the total molecules.[103] Hypophosphorylation of CTD is required for transcription initiation, but hyperphosphorylation during elongation. The CTD consists of a 7-member peptide unit, repeated at least 8 times, with a consensus sequence Y-S-P-S-P-S that is essential for eukayotic cell viability; phosphorylation occurs on serine 2 and 5[104] and ubiquitination occurs subsequently.[103] Many kinases can phosphorylate the CTD, which is a site for numerous protein-protein interactions during transcription elongation and possibly during TCR.

Cells from patients suffering from the neurodegenerative and sun-sensitive disease Cockayne syndrome are specifically defective in TCR.[105,106] DNA and RNA synthesis fails to recover to normal levels after UV irradiation.[107] The excision of DNA photoproducts from total genomic DNA of CS cells is normal, but repair of transcriptionally active genes is reduced.[106] The neurodegenerative symptoms of CS could be a direct result of neuronal loss through apoptosis of nondividing cells that experience natural oxidative damage in the brain. Two CS genes, *CSA* and *CSB*, are involved specifically in TCR. Another gene, tentatively named UV^s after a heterogeneous number of UV sensitive patients with in vitro properties closely resembling CS, may be also involved.[108-111] CSA contains WD-repeat motifs that are important for protein-protein interactions.[112] CSB contains an ATPase, helicase motifs, and a nucleotide binding domain, but only the latter is essential for TCR and CSB does not function as a helicase.[113,114] The CSB protein is associated in the RNA pol II elongation complex with DNA and nascent RNA.[115-117] The CSA protein is neither part of this complex, nor forms a stable complex with CSB,[115] but is still required for TCR during transcription elongation.[116,118] Cells lacking either CSA or CSB are unable to ubiquitinate the CTD of RNA pol II.[119]

In addition, the basal transcription factor, TFIIH, that is involved in transcription initiation and elongation, plays a major role in TCR and many of its components are directly involved in remodeling the damaged regions for excision to occur by both TCR and GGR.[10,120] Mutations in the XPB and XPD components can affect the promoter opening function of TFIIH.[121,122] The XPB and D components of TFIIH and CSB interact with p53,[123,124] suggesting a role for p53 in regulation of TCR. But this role is not as prominent as regulation of GGR in the overall repair reaction.[72,86,87] The XPG gene product, as well as being the 3' nuclease for excision, interacts with RNA pol II and facilitates efficient transcription elongation[125] thereby providing an explanation for the complex symptoms of some XPG patients that show both XP and CS symptoms. XPG is also a cofactor required for the activity of thymine glycosylase (*nth* gene) and is thereby linked to repair of oxidative damage that may also be important in maintaining neural functions that fail in CS.[126] By in vitro reconstruction using recombinant proteins, the complete excision process on metabolically inactive DNA can be carried out with a minimal set of components: RPA, XPA, TFIIH (6 sub-unit core of XPB, XPD, p44, p34, p52, p62, and 3 sub-unit kinase CAK), XPC, hHR23B, XPG, ERCC1 and XPF.[127] CAK appears to act as a negative regulator and inhibition of phosphorylation stimulates repair.

There are several examples of DNA damage that is not subject to repair, but instead interferes with the repair proteins in deleterious fashions. A unique mechanism of cell killing by the tunicate toxin ecteinascidin-743 involves interference specifically with TCR, especially XPG, such that active TCR leads to blocked intermediates that contribute to cell death.[128] Some bulky lesions, especially those from AAF, B[a]P, act as decoys to trap the NER complex and prevent excision.[129] In mixtures of adducts, such as is common from chemical carcinogens and UV, trapping of repair components can result in reduced rates of repair of repairable lesions. The most effective of these "decoy" lesions appears to be a B[a]P adduct opposed by an apyrimidinic site from which the dCMP has been lost.[129] Similarly, 8-oxyguanine can trap DNA topoisomerase in an inactive state.[130]

The Core Excision Process Common to GGR and TCR

After damage recognition through the specific GGR or TCR proteins, a process of damage verification appears to occur in which the initial proteins are replaced by another damage binding complex, XPA and RPA that has a higher binding affinity.[131,132] The individual factors of NER associate sequentially and independently on UV photoproducts without a necessary preassembly of a "repairosome" complex.[120,133-135] The basic components of the process include the XPA protein, which binds directly to the heterotrimeric replication protein (RPA),

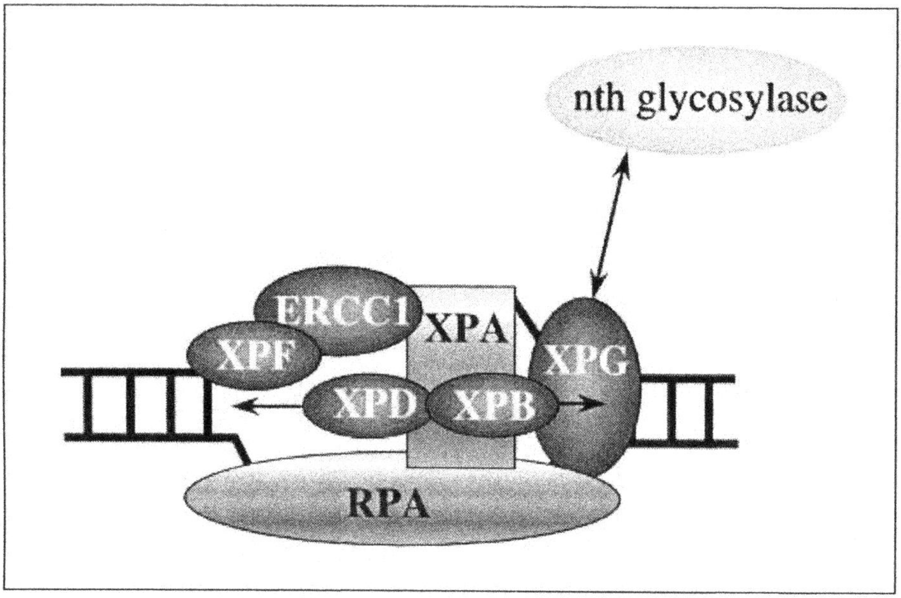

Figure 4. Core excision complex. The UV damage is bound by the XPA/RPA heterodimer, which interacts with the 5'nuclease complex XPF/ERCC1 through the ERCC1 partner. XPB and XPD are helicases that unwind the DNA in opposite directions. XPG is the 3' nuclease that makes the first incision, and interacts with RPA.

the 9 subunit TFIIH, the XPC-hHR23B complex, the XPG nuclease, and the ERCC1-XPF nuclease (Fig. 4).[127,135,136] XPA binds to the p34 subunit of RPA through its N-terminal region and to the p70 subunit through a central portion in exon IV; to ERCC1 through a small region in exon II; to DNA through its central region covering approximately exons III to V and a Zn finger in exon III; and TFIIH through its N-terminal region.[137] TFIIH can interact with all of the components of the core excision complex, individually, but its strongest binding at physiological strengths is with XPC and XPG.[120] The final stable complex that remains at the damaged site is unclear but a subset of these components acts as an assembly point for the XPG and ERCC1/XPF nucleases that cut the damaged strand 3' and 5' to the damaged site.

An in vitro study of the binding activities of various NER components suggests that the XPA-RPA complex is the stronger binding factor on DNA damage.[132] Unwinding of the damaged site involves the 3'-5' (XPB) and 5'-3' (XPD) helicases of TFIIH. The unwound region is stabilized by the XPA binding protein and the RPA heterotrimeric protein that binds to both XPA and DNA. XPA binds to both the damaged and the undamaged strand, but RPA binds mainly to the undamaged strand.[131] In the special case of triplex-induced distortions to DNA, RPA can carry out the primary binding alone, and XPA cannot bind in the absence of RPA, but serves to reduce nonspecific aggregation of RPA.[39]

The excision process involves removal of a 27-29 nt oligonucleotide containing the photoproduct by cleavages 5 nt on the 3' side of the photoproduct, and 24 nt on the 5' side.[138] The structurespecific cleavage pattern is determined by binding of RPA to the unwound damaged site, and the excised fragment is close in size to the footprint of RPA on DNA.[139] Slight variation in the precise sites of cleavage result in the removal of variable fragments between 27 and 29 nt. The XPG nuclease cleaves first on the 3' side of the damage and interacts with the XPC-hHR23B complex and with TFIIH;[120] the XPF-ERCC1 heterodimer cleaves subsequently

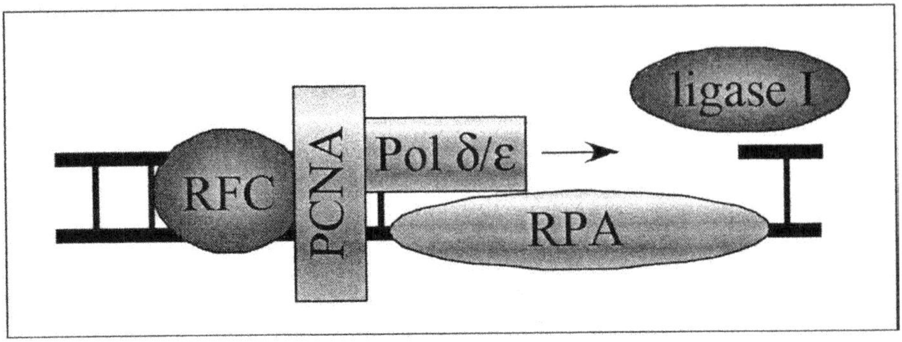

Figure 5. The polymerization reaction. Polymerase D or E (δ, ε) binds to PCNA that is a trimer forming a collar around the DNA, loaded onto DNA by the 5-member complex RFC. The exposed single stranded region is a binding site for RPA, which is a 3-component complex with a footprint on the DNA of similar size to the excision fragment. Ligation occurs by the action of ligase I.

on the 5' side of the damaged site.[66] The XPF and ERCC1 genes may represent an ancient duplication and are unstable unless in a heterodimer complex.[140] XPA serves as an anchor for the 5' nuclease through binding to ERCC1.[141]

The XPG nuclease has a particularly complex range of activities, and mutations in XPG are associated with both XP and CS diseases.[105,142,143] XPG is an endonuclease in the FEN-1 family that is capable of strand-specific cleavage of a range of DNA substrates that may arise during DNA replication, repair and recombination.[144] The protein not only cleaves the 3' side of photoproducts during NER,[145,146] but it also resolves recombination intermediates by strand specific cleavage at double strand-single strand junctions.[144,147] XPG is also a cofactor for thymine glycosylase through which it stimulates repair of oxidative DNA damage (thymine glycols, 8-oxyguanine, etc).[126,146,148] XPG is also involved in transcription elongation by interaction with the elongation factor TFIIS, independently of CSB.[125]

The excised region is replaced by the action of a complex similar to that involved in normal DNA replication (Fig. 5).[127] PCNA is loaded onto the DNA by the 5 subunit RFC complex which then anchors the replicative polymerases Pol D or Pol E.[149] The final closure of the repaired site occurs with DNA ligase, most probably ligase I.[149]

Additional proteins that are not required for the core excision reaction in vitro on synthetic substrates, appear to play a role in vivo in the interaction of repair with other pathways of gene regulation. One protein, XAB2 interacts with XPA, CSA and B, and TFIIH, and may be involved in TCR and transcription.[150] A protein identified in yeast (*S. cerevisiae*) that is responsible for chromatin remodeling, Cnd2, causes a decrease in pyrimidine dimer excision when it is mutated.[151] Nuclear receptors that mediate hormone-dependent gene activation require phosphorylation by the cdk7 component of TFIIH. A novel connection has been demonstrated between hormonal control of gene expression and the *XPD* repair gene in TFIIH, that influences signal transduction by retinoids (RAR and RXR), vitamin D, steroids (ER, AR, and GR) and thyroid hormone.[152] This interaction may also involve the repair gene hMMS19 that interacts with the XPB and D components of TFIIH and is a coactivator of the estrogen receptor involved in RNA pol II transcription.[153,154] The interaction between DNA repair and hormone-dependent pathways may contribute to the complexity of many of the DNA repair deficient human disorders.[4] These interactions may also explain the observation that exposure to the cytokine IL-12,[155] or interferon[156-159] modulates DNA repair.

Ubiquitination and Proteolysis in DNA Repair

Mammalian cells have a major protein degradation pathway by which proteins are marked for destruction through covalent addition of ubiquitin and direction into the 26S proteasome pathway. A series of steps involve the E1 ubiquitin-activating enzyme, the E2 ubiquitin-conjugating enzyme, and the E3 ubiquitin-protein ligases.[160] There are many multimeric E3 ligases that confer specificity on the pathway of protein destruction. The addition of multiple ubiquitin chains to a target protein provides a recognition signal for degradation via the 26S proteasome.[161] Single ubiquitin moeities may be used for modification of protein function, without being sufficient to trigger the proteasome system. The selective destruction of proteins provides a level of control by which cells can switch between successive steps in complex processes. The successive steps of NER, especially TCR, and DNA replication after UV damage may, for example, be controlled by specific degradation of proteins involved in the early steps. For most of the major DNA repair pathways, the damage response involves components that have potential proteolytic functions. Prevention of proteolysis can result either in increased NER [162] or no effect.[163]

The two main components of the GGR pathway, XPC & XPE, involve activation of different proteolytic pathways, and in some interactions the proteasome may regulate GGR without protein degradation. The HR23B partner of the XPC/HR23B damage recognition complex interacts with XPC by its C-terminal end, but its N-terminal region has ubiquitin-like regions (UbL).[164,165] These regions interact with other ubiquitinated proteins and with the 19S proteasome subunit to inhibit protein degradation.[162,165-168] Prevention of proteasome function consequently increases NER capacity.[162]

The DDB2 component of the XPE heterdimer is targeted for degradation by one of the cullin family of E3 ubiquitin ligases. Degradation of p48 occurs as a result of ubiquitination by CUL4A.[169] The functional significance of this interaction for repair is exemplified by the identification of a mutation in the CUL-4A binding site that blocks ubiquitination in a repair-defective XPE patient.[169] Although the p127 component of XPE is also ubiquitinated by CUL4A, the functional significance is more difficult to establish. The p127 protein is normally present in excess, so possibly the small fraction that binds to p48 after irradiation would be regulated differently from the bulk of the protein.

The proteasome system is linked to TCR mainly at the level of clearance of the stalled RNA pol II. Whether proteolytic degradation is part of the TCR process, itself, or is an alternative when TCR is unable to clear lesions ahead of the blocked polymerase is still unclear.[170] Immediately after UV irradiation RNA pol II (large subunit) is hyperphosphorylated[171] and ubiquitinated.[103,119] These processes involve the two CS gene products,[119] and an interaction between the transcription factor TFIIH and SUG1, a component of the 26S proteasome.[172] Since cells lacking CSA or B are unable to ubiquitinate RNA pol II,[119,173] but can hyperphosphorylate RNA pol II[171] the two modifications are either independent or phosphorylation precedes ubiquitination. Phosphorylation itself appears to be sufficient for RNA pol II degradation.[171]

In yeast, however, elimination of the sole ubiquitin ligase, Rsp5, does not change the characteristics of TCR.[163] In yeast, also, CSB forms a complex with another protein, Def1, and *DEF1* mutants can carry out TCR but are unable to degrade RNA pol II after UV irradiation.[174] These results suggest that TCR and RNA pol II degradation may therefore be two alternative pathways of response to transcription arrest.[170,174] There are, however, significant differences between TCR in human and yeast cells. Mutations in the yeast homologs of *CSA* and *CSB* (*Rad 28* and *Rad26*, respectively) do not result in UV sensitive cells, unlike human CS cells, so the details of transcription stalling and RNA pol II clearance may differ in detail.[170]

The overall impression remains that ubiquitination, the proteasome, and proteolysis, are intimately involved in NER and the UV response. But questions remain as to whether proteolytic degradation is part of the process of NER, required for example to clear repair intermediates or RNA pol II from damaged sites, or whether proteolysis is an alternative degradative pathway recruited when repair itself aborts. Recent evidence indicates that the 19S ATPase subunits that constitute the base of the proteasome (Sug1/Rpt6, Sug2/Rpt4, and Yta/Rpt5) are involved with the transcription complex, itself, independently of proteolysis.[175] This recent evidence suggests that monoubiquitination can promote interaction between transcription factors and cognate sequences, but is insufficient to trigger protolysis. This further indicates that the relationship of ubiquitination, the proteasome, proteolysis and NER remain complex.

Inducible Responses to UV Irradiation

Early studies on gene expression after UV damage in bacteria demonstrated a large range of damage-inducible processes. Notably, excision repair and recombinational repair were both induced by a defined biochemical pathway in which repressor cleavage enhanced expression of the genes in the NER and RecA pathway.[1] Numerous attempts were subsequently made to detect induced DNA repair in mammalian cells by dose fractionation or by transfection of unirradiated vectors or viruses into irradiated cells. Although some of these experiments gave results consistent with the enhancement of repair, generally the effects were small and developed slowly over a cell cycle. In contrast, damage did result in induction of genes involved in cell cycle arrest and apoptosis, in many cases through p53. Recent studies using microarrays to identify an unbiassed set of induced genes following irradiation of yeast (*S. cerevisiae*) and human keratinocytes have produced a clearer picture of the transcriptional response. Although the transcription profile of cells does change after damage to DNA, the significance of the changes and their role in the repair of the damage is not obvious.

In *S. cerevisiae* there is now a nearly complete set of deletion mutants for the whole genome that represents an invaluable research tool.[27,28,176] The striking conclusion from these studies is that the genes responsible for sensitivity to a variety of stresses, including high salt, sorbitol, galactose, pH 8, minimal medium and nystatin,[176] and UV radiation[27,28] were not induced by exposure to these conditions. Where a few damage-responsive repair genes were detected, these were no more frequent than expected by chance, when the large number of genes were considered.[27] Direct evidence that induction may not be functionally important in repair was demonstrated by deletion of the damage-responsive promoter elements for the yeast *Rad 2* and *Rad54* genes involved in NER and recombination, respectively.[177,178]

A similar situation has been found in human cells. UV irradiation induced a large number of genes, but most of these represent genes involved in the UV-specific inflammatory response, stress response, basal transcription, splicing, translation, cell cycle arrest and proteasome degradation.[30] Genes that are involved in generalized metabolism and cell-cell adhesion were down-regulated. The only exception to the general lack of induction of repair genes, are the p53 responsive genes *XPC*, *DDB2*, *GADD45*, and *p53* itself, most of which have been discussed earlier in relation to GGR. GADD45 is transcriptionally activated by UV damage, possibly through p53 and Oct1, and also plays a role in GGR through modulation of chromatin structure.[179,180] Low UV doses induce p53 resulting in increased expression of the cell cycle regulator p21, that then interacts with the replication apparatus through PCNA, and down regulation of *HDM2*.[181] High UV doses however increase *GADD45*, *BAX* expression and result in apoptosis.[181] GADD45 may compete with p21 for access to PCNA.[182] Mice that lack *p53* or *GADD45* show increased sensitivity to UV damage and a lack of GGR, but mice that lack *p21* show no change in repair.[183]

The reason for these, perhaps unexpected, observations may be that there is adequate repair capacity under normal conditions to maintain DNA integrity, because of the necessity to repair oxidative damage and damage associated with replication-arrest and other normal cellular processes. Induction may therefore not be needed under most radiation conditions, and instead cells adopt a strategy of alterations in protein function by secondary modification through phosphorylation and ubiquitination.[31-35] Changes in basal levels of expression of repair gene, however, may well occur under a range of conditions of nutrition, hormones, development and differentiation, but these will be different from damage-dependent induction that appears very slight, and restricted to GGR. The induction of expression that does occur from DNA damage may therefore be a cell-wide, general alarm response. The situation may resemble a burglar who breaks one window in a house but the lights and alarms go on all over the house – the response is much larger than the scale of the initial damage and the tools required for repair.

Kinase Activation and Cellular Response to UV Damage

The coordination and regulation of the response of cells to UV damage involves phosphorylation and dephosphorylation of components of the NER system and related checkpoints throughout the cell cycle.[184-186] The complexity of the responses and the bewildering number of primary and secondary kinases is due in part to the different mechanisms of activation that operate as a function of UV dose and cell cycle stage. Numerous different kinases can be activated by UV light, including ATM, ATR, HIP2K, CHK2, and c-Abl, and these act on a large number of substrates. HIPK2 is a newly reported kinase specific for UV, but its substrates and site of action have not been clearly defined, other than p53.[33,34] Activation can proceed through direct interaction between DNA damage and the ATR kinase.[187] Activation of ATR can also occur as a result of DNA replication arrest.[188] Indirect activation of the ATM kinase can occur as a result of NER-induced breaks at low UV doses.[189] Secondary activation of downstream kinases, e.g., CHK2, can proceed as a result of phosphorylation by primary initiator kinases. An important substrate for many of these kinases is p53, resulting in activation of its transactivation function.[31-35,184]

UV light produces effects that can be classified as either nontargeted or targeted. Nontargeted means that the damage activates kinases that then indirectly activate cell cycle checkpoints or DNA repair processes. Targeted means that the damage directly blocks cellular processes such as RNA pol II transcription or progression of DNA replication forks. These direct effects can then give rise to secondary nontargeted effects.[190]

There are a series of functional stages, checkpoints, at which nontargeted kinase activity is important, notably at the transitions between phases of the cell cycle (G1-S, within S, and G2-M). Within the S phase several checkpoints have also been identified: at S phase initiation, at individual replicon initiation within the S phase, and at the actual sites of DNA fork progression.[184,190] Initiation of replication at the G1/S boundary and during S at individual replicons, may be under different regulation than that required for continuation of DNA chain growth.[190-192] The initiation of the S phase and of individual replicons appear to be down-regulated as nontargeted events, that occur as a consequence of signal transduction from initial damage to the actual arrest mechanism, and involves regulatory kinases, p53, and p21. Nontargeted signal transduction results in p53 transactivation leading to p21 binding to PCNA which slows replication fork progression.[192]

Early observations showed that low levels of UV damage, below about 1 $J.m^{-2}$, to normal and ataxia telangiectasia (AT) human cells activated the G1/S checkpoint and inhibited replicon initiation during S.[189,191,193] This was suggestive of nontargeted events analogous to those that occur after Xrays.[184,194] Cells from patients with AT or Nijmegen breakage syndrome (NBS), who have defects in the ATM kinase or the hMre11 complex, are unable to down-regulate

DNA replication after Xray or UV damage.[189,193] But, there are major differences in the Xray and UV signal transduction pathways. P53 and Nbs1 are phosphorylated on different residues after these two kinds of DNA damage, Xrays induce less targeted replication arrest than does UV damage, and ATM/ATR do not seem to be involved after higher doses of UV light.[194]

At higher UV doses, above about 1 $J.m^{-2}$ at which a large fraction of the cell population still survives, targeted arrest of DNA replication forks becomes the dominant response. This is because of the large numbers of photoproducts formed in replicating DNA (more than 1 per replicon) that directly block fork progression.[195,196] Blocked forks may then initiate a unique signal transduction pathway.[190] Replication fork arrest by DNA damage, for example, results in changes in the phosphorylation status of RPA and Brca1[197] which could be starting points for signal transduction. In XP-V cells, which lack the major bypass polymerase, Pol H,[198,199] the targeted arrest is greatly enhanced, but can be over-ridden by adding rhRPA[200] or PCNA[201] to in vitro assays.

The difference between the targeted and nontargeted checkpoints is sharply illustrated by the effects of caffeine. Caffeine increases S-phase arrest after UV damage by increasing the efficiency of the replication block.[16,17,202,203] There is evidence that several kinases of different classes are involved in UV-induced replication arrest, both ATR and nonPI3K.[204] In *S. pombe* caffeine toxicity involves rhp6 (Rad6 in *S. cerevisiae*),[205] which in combination with Rad18 functions as a ubiquitinating enzyme during DNA replication arrest.[206] Caffeine however reverses the G1/S and G2/M checkpoint delays in X-ray damaged cells by inhibiting ATM/ATR and other kinases.[31]

c-Abl is a tyrosine kinase that is activated by Xrays but not UV. Despite this apparent unresponsiveness to UV damage, c-Abl binds to DDB1 and phosphorylates DDB2 reducing its activity.[207] There is a possibility that such tyrosine phosphorylation is signal for the ubiquitination of DDB and its subsequent degradation.[207]

Replication of Damaged DNA

NER can remove DNA damage before DNA replication begins, and consequently plays a major role in reducing the amount of damage that becomes fixed as mutations during replication.[208] DNA photoproducts, however, are blocks to the replicative DNA polymerases, alpha, delta and epsilon (Pol α, δ, ε or A, D, E) which cannot accommodate large distortions such as DNA photoproducts or adducts in their active sites.[209,210] Replicative bypass of these photoproducts is achieved instead by damage-specific polymerases with relaxed substrate specificity, now defined as class Y polymerases (Fig. 6).[211,212]

Three class Y polymerases have been identified in the mammalian genome, pol eta, iota and kappa (POL η, ι, κ, or H, I, and K). The genes *POL H* and *I* are close homologs, unique to mammalian cells, and only a single *POL H* gene is found in yeast.[211] Pol I has a poorer capacity for replication of UV damage and Pol K seems completely unable to replicate UV damage. The apparent base-specificity for Pol H and Pol I indicates that Pol H may exercise a preference for replicating T-containing photoproducts, and Pol I a preference for C-containing sites. Pol H can therefore insert adenines across from thymine-containing photoproducts, resulting in accurate replication of a T-T pyrimidine dimer. Pol I can insert guanines across from cytosine-containing photoproducts, resulting in accurate replication of a T-C pyrimidine dimer or [6-4] photoproduct. These polymerases have larger active sites that allow them to read-through noninformative sequence information resulting from DNA damage.[213] As a consequence these polymerases have high error rates of the order of 1% when assayed in vitro, and this property must be controlled in vivo otherwise the results would be catastrophic to the cell.[214,215]

Control of Pol H activity is achieved by several mechanisms. Pol H acts distributively, and is only able to extend the nascent DNA chain by one or two bases across from the photoproducts, and there may be a role for editing by a separate exonuclease. Chain extension then

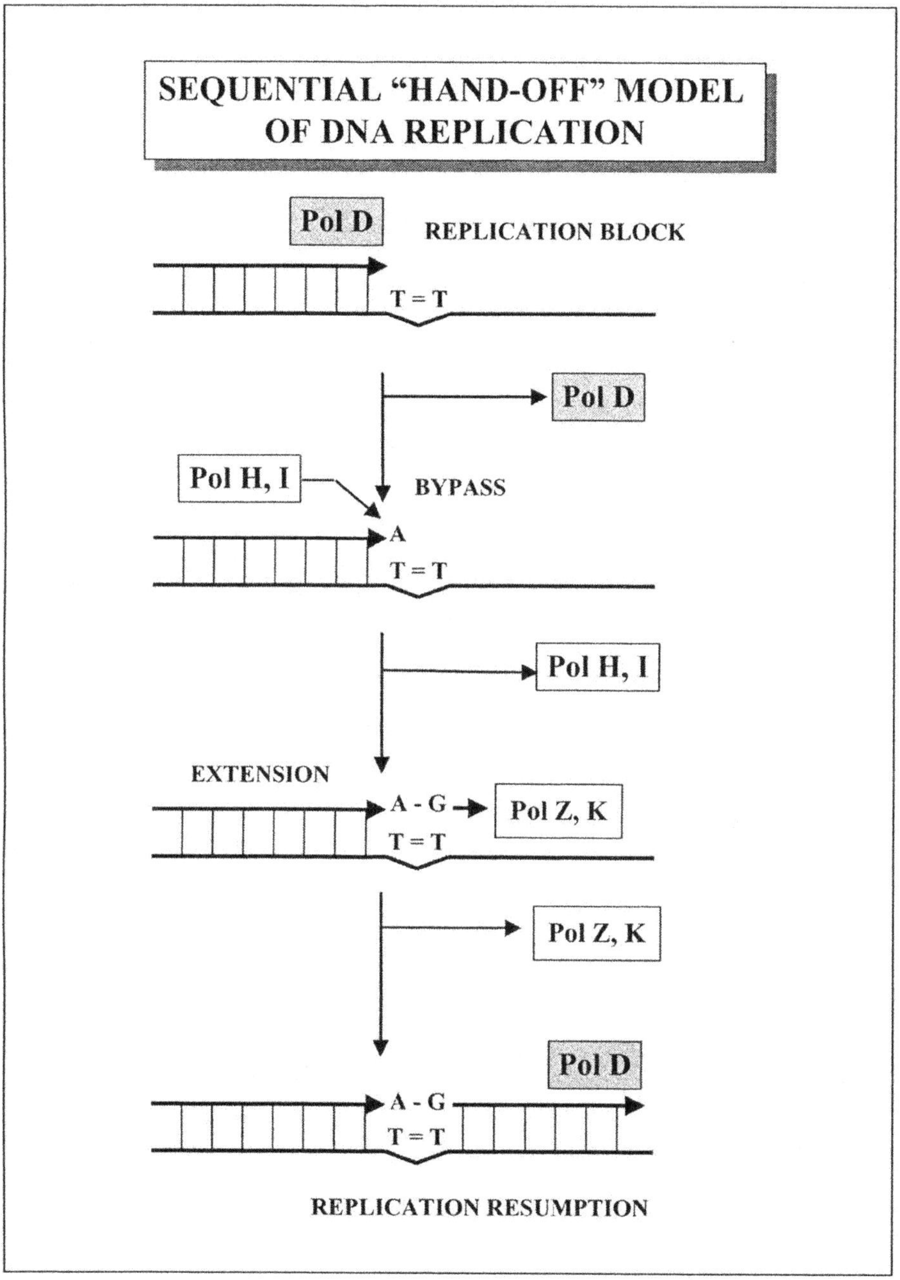

Figure 6. The replication of damaged DNA by the action of low fidelity class Y polymerases, Pol H and I (η, ι). In this model, pol δ is blocked by a damaged base, displaced and replaced by pol η or ι that carries out distributive synthesis of one or 2 bases. These are then replaced by pol ζ or κ, which are high or low fidelity polymerases respectively, that can extend the nascent chain until replaced by pol δ. The many additional proteins involved in replication (e.g., PCNA, RFC and the RAD6 ubiquitinating enzymes) are not shown since their roles are still under study.

requires the activity of polymerase zeta or kappa (pol ζ, κ or Pol Z and K). Pol H is uniformly distributed in the nucleus, and excluded from the replication fork until replication is stalled by UV damage. Pol H then traffics into the nucleus and accumulates with a large number of other proteins in microscopically visible foci at the replication fork.[216,217] This requires specific sequence motifs in the protein for translocation and for binding to PCNA.[216,218]

The error-free DNA replication pathway involving pol H and other polymerases is part of an overall DNA replication pathway for UV damage defined in yeast as the RAD 6 epistasis group[1] Several of the components of this pathway are ubiquitinating enzymes that target PCNA, which functions as a major control point for the overall process.[219] PCNA is modified by UBC9 that is a small ubiquitin-related modifying enzyme (SUMO), and by Rad6 and the MMS2-UBC13 heterodimer that are ubiquitin-conjugating enzymes.[219] Multiple modifications compete for the same lysine (K164) residue on PCNA and redirect subsequent pathways through SUMO, mono and multi-ubiquitination.[219]

In the absence of the low fidelity polymerases, a complex set of overlapping intra-S phase checkpoints are activated that prevent gross chromosome rearrangements.[190,220,221] Blocked replication forks are further stabilized in UV damaged cells by p53, and its inactivation by viral transformation allows fork breakage and DNA rearrangements to occur.[26,220-222] A central player in these rearrangements is the protein complex hMre11/hRad50/Nbs1 that is involved in homologous and nonhomologous recombination, DNA replication arrest and telomere maintenance.[190,220,221,223] The hMre11 complex is phosphorylated on its hMre11 and Nbs1 subunits by ATM, ATR and other kinases, according to the nature of the DNA damage.[190,223]

Differentiation and Tissue-Specific Expression of DNA Repair Genes

One of the expanding areas of interest in DNA repair is the extent to which it is subject to changes in the functional, differentiated or malignant state of the cell. There is a general consensus emerging that as cells differentiate to a nondividing state their capacity for NER of UV damage declines. The major loss is in the capacity for repair of transcriptionally inactive genes or regions of the genome.[224-228] In keratinocytes although NER is dependent on p53, after differentiation there is also a p53-independent reduction in NER.[228] This phenomenon still requires a detailed understanding of the molecular events associated with the decline: does it involve a few specific genes, such as DDB1/2 or XPC, or does it represent a more general phenomenon involving nucleotide precursor pools or chromatin, for example? A change in chromatin conformation due to mutations in the *Cd2* gene in yeast results in the loss of pyrimidine dimer excision.[151]

More specific down-regulation of repair has been described in the testis where a reduction in NER is due to reduced expression of XPA, rendering testicular tumors more sensitive to cis-PT chemotherapy.[229,230] In the basal layer of the skin, a small fraction of cells seem to be specifically defective in repair of UV and carcinogen damage and retain photoproducts for long times after irradiation.[231] The mechanism for this will be particularly interesting because it represents down-regulation of repair in a few cells distributed among a large population of repair-competent cells.

Polymorphisms in DNA Repair Genes

The disease alleles of the NER genes represented in the human diseases XP and CS and others represent high penetrance, causative mutations.[75] There are in addition susceptibility alleles that may give rise to disease at lower penetrance according to life-style, diet, hormonal balance, and exposure to environmental chemical, radiation and infectious agents. Polymorphisms that are not directly associated with the XP phenotype have been described for many of the NER and other repair genes.[232] Polymorphisms have been found in the XPA, XPC and D genes.[233] Many of these polymorphisms result in nonconservative amino acid changes that could have an impact on NER function and disease.[234] A large number of alleles present at low frequency individually could combine to create variation in disease frequency in the overall population.

Acknowledgements

I was supported during the writing of this review by the National Institutes of Environmental Health Sciences grant 1 RO1 ES 8061, the Ellison Senior Scholar Award, and the Cancer Center Support Grant from NCI. I am also very grateful to the XP Society, Poughkeepsie, NY, the Luke O'Brien Fund for Cockayne Syndrome, and the Marvin and Roma Auerback fund for melanoma research for their continued support and encouragement.

References

1. Friedberg EC, Walker G, Siede W. DNA Repair and Mutagenesis Washington DC: ASM Press, 1995.
2. Wood RD, Mitchell M, Sgouros J et al. Human DNA repair genes. Science 2001; 291:1284-1289.
3. Cleaver JE, Crowley E. UV damage, DNA repair and skin carcinogenesis. Front Biosci 2002; 7:d1024-1043.
4. Bootsma D, Kraemer KH, Cleaver JE et al. Nucleotide excision repair syndromes: Xeroderma pigmentosum, Cockayne syndrome, and trichothiodystrophy. In: Vogelstein B, Kinzler KW, eds. The Genetic Basis of Human Cancer. McGraw-Hill, 1998:245-274.
5. Thompson LH. Nucleotide excision repair: Its relation to human disease. In: Nickoloff JA, Hoekstra M, eds. DNA Repair in Higher Eukaryotes. Totowa: Humana Press, 1998:335-393.
6. Kraemer KH, Lee MM, Scotto J. Xeroderma pigmentosum: Cutaneous, ocular and neurological abnormalities in 830 published cases. Arch Dermatol 1987; 123:241-250.
7. Cleaver JE. DNA repair and its coupling to DNA replication in eukaryotic cells. Biochim Biophys Acta 1978; 516:489-516.
8. Hanawalt PC. Transcription-coupled repair and human disease. Science 1994; 266:1957-1958.
9. Bohr VA. Gene specific DNA repair. Carcinogen. 1991; 12:1983-1992.
10. Schaeffer L, Roy R, Humbert S et al. DNA repair helicase: A component of BTF2 (TFIIH) basic transcription factor. Science 1993; 260:58-63.
11. Wang Y, Cortez D, Yazdi P et al. BASC, a super complex of BRCA1-Associated proteins involved in the recognition and repair of aberrant DNA structures. Genes Dev 2000; 14:927-939.
12. Cleaver JE. Common pathways for ultraviolet carcinogenesis in the repair and replication defective groups of xeroderma pigmentosum. J Dermatol Sci 2000; 23:1-11.
13. Cleaver JE, Kaufmann WK, Kapp LN et al. Replicon size and excision repair as factors in the inhibition and recovery of DNA synthesis from ultraviolet damage. Biochim Biophys Acta 1983; 739:207-215.
14. Cleaver JE. Normal reconstruction of DNA supercoiling and chromatin structure in Cockayne syndrome cells during repair of damage from ultraviolet leight. American Journal of Human Genetics 1982; 34:566-575.
15. Domon M, Rauth AM. Ultraviolet irradiation of mouse L cells: Effects on DNA synthesis and progression through the cell cycle. Radiat Res 1968; 35:350-368.
16. Domon M, Rauth AM. Effects of caffeine on ultraviolet irradiated mouse L cells. Radiat Res 1969; 39:207-221.
17. Cleaver JE. Caffeine toxicity is inversely related to DNA repair in simian virus-40 transformed xeroderma pigmentosum cells irradiated with ultraviolet light. Teratogen Carcinogen Mutagen 1989; 9:147-155.
18. Kolodner R. Biochemistry and genetics of eukaryotic mismatch repair. Genes Dev 1996; 10:1433-1442.

19. Kolodner RD, Marsischky GT. Eukaryotic DNA mismatch repair. Curr Opin Genet Dev 1999; 9:89-96.
20. Marra G, Schar P. Recognition of DNA alterations by the mismatch repair system. Biochem J 1999; 338(Pt 1):1-13.
21. Mellon I, Bohr VM, Hanawalt PC. Preferential repair of an active gene in human cells. Proc Natl Acad Sci USA 1986; 83:8878-8882.
22. Leadon SA, Avrutskaya AV. Differential involvement of the human mismatch repair proteins, hMLH1 and hMSH2, in transcription-coupled repair. Cancer Res 1997; 57:3784-3791.
23. Leadon SA, Avrutskaya AV. Requirement for DNA mismatch repair proteins in the transcription-coupled repair of thymine glycols in Saccharomyces cerevisiae. Mutat Res 1998; 407:177-187.
24. Bellacosa A. Functional interactions and signaling properties of mammalian DNA mismatch repair proteins. Cell Death Diff 2001; 8:1076-1092.
25. Ljungman M, Zhang F. Blockage of RNA polymerase as a possible trigger for u.v. light-induced apoptosis. Oncogene 1996; 13:823-831.
26. Cleaver JE, Bartholomew J, Char S et al. Polymerase η and p53 jointly regulate cell survival, apoptosis and Mre11 recombinaiton during S phase checkpoint arrest after UV irradiation. DNA Repair 2001; 3:1-17.
27. Birrell GW, Brown JA, Wu HI et al. Transcriptional response of Saccharomyces cerevisiae to DNA-damaging agents does not identify the genes that protect against these agents. Proc Natl Acad Sci USA 2002; 99:8778-8783.
28. Birrell GW, Giaever G, Chu AM et al. A genome-wide screen in Saccharomyces cerevisiae for genes affecting UV radiation sensitivity. Proc Natl Acad Sci USA. 2001; 98:12608-12613.
29. Hoh J, Jin S, Parrado T et al. The p53MH algorithm and its application in detecting p53-responsive genes. Proc Natl Acad Sci USA 2002; 99:8467-8472.
30. Sesto A, Navarro M, Burslem F et al. Analysis of the ultraviolet B response in primary human keratinocytes using oligonucleotide microarrays. Proc Natl Acad Sci USA 2002; 99:2965-2970.
31. Sarkaria JN, Busby EC, Tibbetts RS et al. Inhibition of ATM and ATR kinase activities by the radiosensitizing agent, caffeine. Cancer Res 1999; 59:4375-4382.
32. Shiloh Y. ATM and ATR: Networking cellular responses to DNA damage. Curr Opin Genet Dev 2001; 11:71-77.
33. D'Orazi G, Cecchinelli B, Bruno T et al. Homeodomain-interacting protein kinase-2 phosphorylates p53 at Ser 46 and mediates apoptosis. Nature Cell Biol 2002; 4:11-19.
34. Hofmann TG, Moller A, Sirma H et al. Regulation of p53 activity by its interaction with homeodomain-interacting protein kinase-2. Nature Cell Biol 2002; 4:1-10.
35. Chehab NH, Malikzay A, Appel M et al. Chk2/hCds1 functions as a DNA damage checkpoint in G1 by stabilizing p53. Genes Dev 2000; 14:278-288.
36. Hosfield DJ, Daniels DS, Mol CD et al. DNA damage recognition and repair pathway coordination revealed by the structural biochemistry of DNA repair enzymes. Prog Nucleic Acid Res Mol Biol 2001; 68:315-347.
37. Cleaver JE, Mitchell DL. Ultraviolet Radiation Carcinogenesis. In: Holland JF, Bast Jr RC, Morton DL et al., eds. Cancer Medicine. Baltimore: Williams and Wilkins, 1997:307-318.
38. Kuraoka I, Bender C, Romieu A et al. Removal of oxygen free-radical-induced 5',8-purine cyclodeoxynucleosides from DNA by the nucleotide excision-repair pathway in human cells. Proc Natl Acad Sci USA 2000; 97:3832-3837.
39. Vasquez KM, Christensen J, Li L et al. Human XPA and RPA DNA repair proteins participate in specific recognition of triplex-induced helical distortions. Proc Natl Acad Sci USA 2002; 99:5848-5853.
40. Cook JS, McGrath JR. Photoreactivating-enzyme activity in metazoa. Proc Natl Acad Sci USA 1967; 58:1359-1365.
41. Sancar A. Structure and function of DNA photolyase. Biochemistry 1994; 33:2-9.
42. Cleaver JE. Photoreactivation: A radiation repair mechanism absent from mammalian cells. Biochem Biophys Res Comm 1966; 24:569-576.

43. Mortelmans K, Cleaver JE, Friedberg EC et al. Photoreactivation of thymine dimers in UV-irradiated human cells: Unique dependence on the culture conditions. Mutat Res 1977; 44:433-446.
44. Hsu DS, Zhao X, Zhao S et al. Putative human blue-light photoreceptors hCRY1 and hCRY2 are flavoproteins. Biochemistry 1996; 35:13871-13877.
45. Ward JF. The complexity of DNA damage: Relevance to biological consequences. Int J Radiat Biol 1994; 66:427-432.
46. Wang H, Zeng ZC, Bui TA et al. Nonhomologous end-joining of ionizing radiation-induced DNA double-stranded breaks in human tumor cells deficient in BRCA1 or BRCA2. Cancer Res 2001; 61:270-277.
47. Cappelli E, Taylor R, Cevasco M et al. Involvement of XRCC1 and DNA ligase III gene products in DNA base excision repair. J Biol Chem 1997; 272:23970-23975.
48. Thompson LH. Evidence that mammalian cells possess homologous recombination mechanisms. Mutat Res 1996; 363:77-88.
49. Iliakis G, Metzher L, Muschel RJ et al. Induction and repair of DNA double strand breaks in radiation-resistant cells obtained by transformation of primary rat embryo cells with the oncogenes H-ras and V-myc. Radiat Res 1990; 50:6575-6579.
50. Jeggo PA, Taccioli GE, Jackson SP. Menage a trois: Double-strand break repair, V(D)J recombination and DNA-PK. BioEssays 1995; 17:949-957.
51. Takata M, Sasaki MS, Sonoda E et al. Homologous recombination and nonhomologous endjoining pathways of vertebrate cells have overlapping roles in DNA double-strand breaks repair and maintenance of chromosomal integrity. EMBO J 1998 in press.
52. Masson M, Niedergang C, Schreiber V et al. XRCC1 is specifically associated with poly (ADP-ribose) polymerase and negatively regulates its activity following DNA damage. Mol Cell Biol 1998; 18:3563-3571.
53. Wei YF, Carter KC, Wang RP et al. Molecular cloning and functional analysis of a human cDNA encoding an Escherichia coli AlkB homolog, a protein involved in DNA alkylation damage repair. Nucleic Acids Res 1996; 24:931-937.
54. Trewick SC, Henshaw TF, Hausinger RP et al. Oxidative demethylation by Escherichia coli AlkB directly reverts DNA base damage. Nature 2002; 419:174-178.
55. Falnes PO, Johansen RF, Seeberg E. AlkB-mediated oxidative demethylation reverses DNA damage in Escherichia coli. Nature 2002; 419:178-182.
56. Shivji MK, Eker AP, Wood RD. DNA repair defect in xeroderma pigmentosum group C and complementing factor from HeLa cells. J Biol Chem 1994; 269:22749-22757.
57. Masutani C, Sugasawa K, Yanagisawa J et al. Purification and cloning of a nucleotide excision repair complex involving the xeroderma pigmentosum group C protein and a human homologue of yeast RAD23. EMBO J 1994; 13:1831-1843.
58. Sugasawa K, Ng JMY, Masutani C et al. Xeroderma pigmentosum group C protein complex is the initiator of global nucleotide excision repair. Mol Cell 1998; 2:223-232.
59. Wood RD. DNA damage recognition during nucleotide excision repair in mammalian cells. Biochimie 1999; 81:39-44.
60. Baxter BK, Smerdon MJ. Nucleosome unfolding during DNA repair in normal and xeroderma pigmentosum (group C) human cells. J Biol Chem 1998; 273:17517-17524.
61. Mansbridge JN, Hanawalt PC. Domain-limited repair of DNA in ultraviolet irradiated fibroblasts from xeroderma pigmentosum complementaiton C. In: Friedberg EC, Bridges BR, eds. Cellular responses to DNA damage. UCLA Symposium on molecular and cellular biology, New series. New York: Alan R. Liss, 1983:195-207.
62. Karentz D, Cleaver JE. Excision repair in xeroderma pigmentosum group C but not group D is clustered in a small fraction of the total genome. Mutat Res 1986; 165:165-174.
63. Kantor GJ, Barsalou LS, Hanawalt PC. Selective repair of specific chromatin domains in UV-irradiated cells from xeroderma pigmentosum. Mutat Res DNA Repair 1990; 235:171-180.
64. Surralles J, Ramirez MJ, Marcos R et al. Clusters of transcription-coupled repair in the human genome. Proc Natl Acad Sci USA 2002; 99:10571-10574.

65. Araki M, Masutani C, Takemura M et al. Centorsome protein centrin 2/caltractin 1 is part of the xeroderma pigmentosum group C complex that initiates global nucleotide excision repair. J Biol Chem 2001; 276:18665-18672.
66. Batty DP, Wood RW. Damage recognition in nucleotide excision repair of DNA. Gene 2000; 241:193-204.
67. Mu D, Sancar A. Model for XPC-independent transcription-coupled repair of pyriomidine dimers. J Biol Chem 1997; 272:7570-7573.
68. Mu D, Hsu DS, Sancar A. Reaction-mechanism of human DNA-repair excision nuclease. J Biol Chem 1996; 271:8285-8294.
69. Itoh T, Linn S, Ono T et al. Reinvestigation of the classification of five cell strains of xeroderma pigmentosum group E with reclassification of three of them. J Invest Dermatol 2000; 114:1022-1029.
70. Chu G, Chang E. Xeroderma pigmentosum group E cells lack a nuclear factor that binds to damaged DNA. Science 1988; 242:564-567.
71. Keeney S, Chang GJ, Linn S. Characterization of a human DNA damage binding protein implicated in xeroderma pigmentosum E. J Biol Chem 1993; 268:21293-21300.
72. Nichols AF, itoh T, Graham JA et al. Human damage-specific DNA binding protein p48. Characterization of XPE mutations and regulation following UV irradiation. J Biol Chem 2001; 275:21422-21428.
73. Hwang BJ, Toering S, Francke U et al. p48 Activates a UV-damaged-DNA binding factor and is defective in xeroderma pigmentosum group E cells that lack binding activity. Mol Cell Biol 1998; 18:4391-4399.
74. Nichols AF, Ong P, Linn S. Mutations specific to the xeroderma pigmentosum group E Ddb- phenotype. J Biol Chem 1996; 271:24317-24320.
75. Cleaver JE, Thompson LH, Richardson AS et al. A summary of mutations in the UV-sensitive disorders: Xeroderma pigmentosum, Cockayne syndrome, and trichothiodystrophy. Hum Mutat 1999; 14:9-22.
76. Itoh T, Linn S. XP43TO, Previously Classified as Xeroderma Pigmentosum Group E, Should Be Reclassified as Xeroderma Pigmentosum Variant. J Invest Dermatol 2001; 117:1672-1674.
77. Reardon JT, Nichols AF, Keeney S et al. Comparative analysis of binding of human damaged DNA-binding protein (XPE) and Escherichia coli damage recognition binding protein (UvrA) to the major ultraviolet photoproducts: T[c,s]T, T[t,s]T, T[6-4]T, and T[Dewar]T. J Biol Chem 1993; 268:21301-21308.
78. Aboussekhra A, Biggerstaff M, Shivji MKK et al. Mammalian DNA nucleotide excision repair reconstituted with purified protein components. Cell 1995; 80:859-868.
79. Bessho T, Sancar A, Thompson LH et al. Reconstitution of human excision nuclease with recombinant XPF-ERCC1 complex. J Biol Chem 1997; 272:3833-3837.
80. Mu D, Park CH, Matsunaga T et al. Reconstitution of human DNA repair excision nuclease in a highly defined system. J Biol Chem 1995; 270:2415-2418.
81. Kazantsev A, Mu D, Nichols AF et al. Functional complementation of xeroderma pigmentosum complementation group E by replication protein A in an invitro system. Proc Am Acad Sci USA 1996; 93:5014-5018.
82. Rapic Otrin V, Kuraoka I, Nardo T et al. Relationship of the xeroderma pigmentosum group E DNA repair defect to the chromatin and DNA binding proteins UV-DDB and replication protein A. Mol Cell Biol 1998; 18:3182-3190.
83. Tang JY, Hwang BJ, Ford JM et al. Xeroderma pigmentosum p48 gene enhances global genomic repair and suppresses UV-induced mutagenesis. Mol Cell 2000; 5:737-744.
84. Liu W, Nichols AF, Graham JA et al. Nuclear transport of human DDB protein induced by ultraviolet light. J Biol Chem 2000; 275:21429-21434.
85. Shiyanov P, Hayes SA, Donepudi M et al. The naturally occurring mutants of DDB are impaired in stimulating nuclear import of the p125 subunit and E2F1-activated transcription. Mol Cell Biol 1999; 19:4935-4943.
86. Tan T, Chu G. p53 Binds and activates the xeroderma pigmentosum DDB2 gene in humans but not mice. Mol Cell Biol 2002; 22:3247-3254.

87. Hwang BJ, Ford JM, Hanawalt PC et al. Expression of the p48 xeroderma pigmentosum gene is p53-dependent and is involved in global genome repair. Proc Am Acad Sci USA 1999; 96:424-8.
88. Trosko JE, Kasschau MR. Study of pyrimidine dimers in mammalian cells surviving low doses of ultraviolet radiation. Photochem Photobiol 1967; 6:215-219.
89. Painter RB, Cleaver JE. Repair replication, unscheduled DNA synthesis, and the repair of mammalian DNA. Radiat Res 1969; 37:451-466.
90. Klimek M. Thymine dimerization in L-strain mammalian cells after irradiation with ultraviolet light and the search for repair mechanisms. Photochem Photobiol 1966; 5:603-607.
91. Hayes S, Shiyanov P, Chen X et al. DDB, a putative DNA repair protein, can function as a transcriptional partner of E2F1. Mol Cell Biol 1998; 18:240-249.
92. Lee TH, Elledge SJ, Butel JS. Hepatitis B virus X protein interacts with a probable cellular DNA repair protein. J Virol 1995; 69:1107-1114.
93. Lin GY, Paterson RG, Richardson CD et al. The V protein of the paramyxovirus SV5 interacts with damage-specific DNA binding protein. Virology 1998; 249:189-2000.
94. Watanabe T, Sukegawa J, Sukegawa I et al. A 127-kDa protein (UV-DDB) binds to the cytoplasmic domain of the Alzheimer's amyloid precursor protein. J Neurochem 1999; 72:549-556.
95. Mellon I, Spivak G, Hanawalt PC. Selective removal of transcription-blocking DNA damage from the transcribed strand of the mammalian DHFR gene. Cell 1987; 51:241-249.
96. Mellon I, Rajpal DK, Koi M et al. Transcription-coupled repair deficiency and mutations in mismatch repair genes. Science 1996; 272:557-560.
97. Gao S, Drouin R, Holmquist GP. DNA repair rates mapped along the human PGK-1 gene at nucleotide resolution. Science 1994; 263:1438-1440.
98. Gowen LC, Avrutskaya AV, Latour AM et al. BRCA1 required for transcription-coupled repair of oxidative damage. Science 1998; 281:1009-1012.
99. Abbott DW, Thompson ME, Robinson-Benion C et al. BRCA1 expression restores radiation resistance in BRCA1-defective cancer cells through enhancement of transcription-coupled DNA repair. J Biol Chem 1999; 274:18802-18812.
100. Sweder KS, Verhage RA, Crowley DJ et al. Mismatch repair mutants in yeast are not defective in transcription-coupled DNA repair of UV-induced DNA damage. Genetics 1996; 143:1127-1135.
101. Sonneveld E, Vrieling H, Mullenders LH et al. Mouse mismatch repair gene Msh2 is not essential for transcription-coupled repair of UV-induced cyclobutane pyrimidine dimers. Oncogene 2001; 20:538-541.
102. Tanaka K, Kamiuchi S, Ren Y et al. UV-induced skin carcinogenesis in xeroderma pigmentosum group A (XPA) gene-knockout mice with nucleotide excision repair-deficiency. Mutat Res 2001; 477:31-40.
103. Lee KB, Wang D, Lippard SJ et al. Transcription-coupled repair and DNA damage-dependent ubiquitination of RNA polymerase II in vitro. Proc Natl Acad Sci USA 2002; 99:4239-4244.
104. Stiller JW, Hall BD. Evolution of the RNA polymerase II C-terminal domain. Proc Natl Acad Sci USA. 2002; 99:6091-6096.
105. Nance MA, Berry SA. Cockayne syndrome: Review of 140 cases. Amer J Med Genet 1992; 42:68-84.
106. Venema J, Mullenders JH, Natarajan AT et al. The genetic defect in Cockyne syndrome is associated with a defect in repair of UV-induced DNA damage in transcriptionally active DNA. Proc Am Acad Sci USA 1990; 87:4707-4711.
107. Lehmann AR, Kirk-Bell S, Mayne L. Abnormal kinetics of DNA synthesis in ultraviolet light-irradiated cells from patients with Cockayne syndrome. Cancer Res 1979; 39:4237-4241.
108. Cleaver JE, Charles WC, McDowell M et al. Are eight xeroderma pigmentosum groups (A-G,V) and two Cockayne syndrome groups (A,B) the whole story in DNA repair? In: Bohr VA, Wasserman K, Kraemer KH, eds. DNA repair mechanisms. Copenhagen: Munksgaard, 1992:56-67.
109. Fujiwara Y, Ichihashi M, Kano Y et al. A new human photosensitive subject with a defect in the recovery of DNA synthesis after ultraviolet-light irradiation. J Invest Dermatol 1981; 77:256-263.
110. Itoh T, Ono T, Yamaizumi M. A new UV-sensitive syndrome not belonging to xeroderma pigmentosum or Cockayne syndrome: Siblings showing biochemical characteristics of Cockayne syndrome without typical clinical manifestations. Mutat Res 1994; 314:233-248.

111. Itoh T, Fujiwara Y, Ono T et al. UVs syndrome, a new general category of photosensitive disorder with defective DNA repair, is distinct from xeroderma pigmentosum variant and rodent complementation group I [see comments]. Am J Hum Genet 1995; 56:1267-1276.
112. Henning KA, Li L, Iyer N et al. The Cockayne syndrome group A gene encodes a WD repeat protein that interacts with the CSB protein and a subunit of RNA polymerase II, TFIIH. Cell 1995; 82:555-564.
113. Citterio E, Rademakers S, van der Horst GT et al. Biochemical and biological characterization of wild-type and ATPase- deficient Cockayne syndrome B repair protein. J Biol Chem 1998; 273:11844-11851.
114. Selby CP, Sancar A. Human transcription-repair coupling factor CSB/ERCC6 is a DNA- Stimulated ATPase but is not a helicase and does not disrupt the ternary transcription complex of stalled RNA polymerase II. J Biol Chem 1997; 272:1885-1890.
115. van Gool AJ, Citterio E, Rademakers S et al. The Cockayne syndrome B protein, involved in transcription-Coupled DNA repair, resides in an RNA polymerase II-containing complex. EMBO J 1997; 16:5955-5965.
116. Tantin D, Kansal A, Carey M. Recruitment of the putative repair coupling factor CSB/ERCC6 to RNA polymerase elongation complex. Mol Cell Biol 1997; 17:6803-6814.
117. Selby CP, Sancar A. Cockayne syndrome group B protein enhances elongation by RNA polymerase II. Proc Natl Acad Sci USA 1997; 94:11205-11209.
118. Tu Y, Bates S, Pfeiffer GP. The transcription-repair coupling factor CSA is required for efficent repair only during the elongation stages of RNA polymerase II transcription. Mutat Res 1998; 400:143-151.
119. Bregman DB, Halaban R, van Gool AJ et al. UV-induced ubiquitination of RNA polymerase II: A novel modification deficient in Cockayne syndrome cells. Proc Natl Acad Sci USA 1996; 93:11586-11590.
120. Araujo SJ, Nigg EA, Wood RD. Strong functional interactions of TFIIH with XPC and XPG in human DNA nucleotide excision repair, without a preassembled repairosome. Mol Cell Biol 2001; 21:2281-2291.
121. Coin F, Marinoni JC, Rodolfo C et al. Mutations in the XPD helicase gene result in XP and TTD phenotypes, preventing interaction between XPD and the p44 subunit of TFIIH. Nature Genet 1998; 20:184-188.
122. Coin F, Bergmann E, Tremeau-Bravard A et al. Mutations in XPB and XPD helicases found in xeroderma pigmentosum patients impair the transcription function of TFIIH. EMBO J 1999; 18:1357-1366.
123. Wang XW, Yeh H, Schaeffer L et al. p53 modulation of TFIIH-associated nucleotide excision repair activity. Nature Genet 1995; 10:188-195.
124. Wang XW, Vermeulen W, Coursen JD et al. The XPB and XPD DNA helicases are components of the p53-mediated apoptosis pathway. Genes Dev 1996; 10:1219-1232.
125. Lee S-K, Yu S-L, Prakash L et al. Requirement of yeast RAD2, a homolog of human XPG gene,for efficient RNA polymerase II transcription: Implications for Cockayne syndrome. Cell 2002; 109:823-834.
126. Cooper PK, Nouspikel T, Clarkson SG et al. Defective transcription coupled repair of oxidative base damage in Cockayne syndrome patients from XP group G. Science 1997; 275:990-993.
127. Araujo SJ, Tirode F, Coin F et al. Nucleotide excision repair of DNA with recombinant human proteins: Definition of the minimal set of factors, active forms of TFIIH, and modulation by CAK. Genes Dev 2000; 14:349-359.
128. Takebayashi Y, Pourquier P, Zimonjic DB et al. Antiproliferative activity of ecteinascidin 743 is dependent upon transcription-coupled nucleotide-excision repair. Nat Med 2001; 7:961-966.
129. Buterin T, Hess MT, Gunz D et al. Trapping of DNA nucleotide excision repair factors by nonrepairable carcinogen adducts. Cancer Res 2002; 62:4229-4235.
130. Lesher DT, Pommier Y, Stewart L et al. 8-Oxoguanine rearranges the active site of human topoisomerase I. Proc Natl Acad Sci USA 2002; 99:12102-12107.

131. Hermanson-Miller IL, Turchi JJ. Strand-specific binding of RPA and XPA to damaged duplex DNA. Biochemistry 2002; 41:2402-2408.
132. Wakasugi M, Sancar A. Order of assembly of human DNA repair excision nuclease. J Biol Chem 1999; 274:18759-18768.
133. Houtsmuller AB, Rademakers S, Nigg AL et al. Action of DNA repair endonuclease ERCC1/XPF in living cells. Science 1999; 284:958-961.
134. Volker M, Mone MJ, Karmakar P et al. Sequential assembly of the nucleotide excision repair factors in vivo. Mol Cell 2001; 8:213-224.
135. Wakasugi M, Sancar A. Assembly, subunit composition, and footprint of human DNA repair excision nuclease. Proc Natl Acad Sci USA 1998; 95:6669-6674.
136. Sancar A. Mechanisms of DNA excision repair. Science 1994; 266:1954-1956.
137. States JC, McDuffie ER, Myrand SP et al. Distribution of mutations in the human xeroderma pigmentosum group A gene and their relationships to the functional regions of the DNA damage recognition protein. Hum Mutat 1998; 12:103-113.
138. Huang JC, Sancar A. Determination of minimum substrate size for human excinuclease. J Biol Chem 1994; 269:19034-19044.
139. Matsunaga T, Park CH, Bessho T et al. Replication protein A confers structurespecific endonuclease activities to the XPF-ERCC1 and XPG subunits of human DNA repair excision nuclease. J Biol Chem 1996; 271:11047-11050.
140. Gaillard PH, Wood RD. Activity of individual ERCC1 and XPF subunits in DNA nucleotide excision repair. Nucleic Acids Research 2001; 29:872-879.
141. Saijo M, Kuraoka I, Masutani C et al. Sequential binding of DNA repair proteins RPA and ERCC1 to XPA in vitro. Nucleic Acids Res 1996; 24:4719-4724.
142. Nouspikel T, Clarkson SG. Mutations that disable the DNA repair gene XPG in a xeroderma pigmentosum group G patient. Hum Mol Genet 1994; 3:963-967.
143. Nouspikel T, Lalle P, Leadon SA et al. A common mutational pattern in Cockayne syndrome patients from xeroderma pigmentosum group G: Implications for a second XPG function. Proc Natl Acad Sci USA 1997; 94:3116-3121.
144. Lieber MR. The FEN-1 family of structurespecific nucleases in eukaryotic DNA replication, recombination and repair. Bioessays 1997; 19:233-240.
145. O'Donovan A, Davies AA, Moggs JG et al. XPG endonuclease makes the 3' incision in human DNA nucleotide excision repair. Nature 1994; 371:432-435.
146. LePage F, Kuroh EE, Aurutskaya A et al. Transcription-coupled repair of 8-oxoGuanine: Requirement for XPG, TFIIH, and CSB and implications for Cockayne syndrome. Cell 2000; 101:159-171.
147. Davies AA, Friedberg EC, Tomkinson AE et al. Role of the Rad1 and Rad10 proteins in nucleotide excision repair and recombination. J Biol Chem 1995; 270:24638-24641.
148. Klungland A, Hoss M, Gunz D et al. Base excision repair of oxidative DNA damage activated by the XPG protein. Mol Cell 1999; 3:33-42.
149. Wood RD. Nucleotide excision repair in mammalian cells. J Biol Chem 1997; 272:23465-23468.
150. Nakatsu Y, Asahina H, Citterio E et al. XAB2, a novel tetratricopeptide repeat protein involved in transcription-coupled DNA repair and transcription. J Biol Chem 2000; 275:34931-34937.
151. Aono N, Sutani T, Tomonaga T et al. Cnd2 has dual roles in mitotic condensation and interphase. Nature 2002; 417:197-202.
152. Keriel A, Stary A, Sarasin A et al. XPD mutations prevent TFIIH-dependent transactivation by nuclear receptors and phosphorylation of RARα. Cell 2002; 109:125-135.
153. Wu X, Li H, Chen JD. The human homologue of the yeast DNA repair and TFIIH regulator MMS19 is an AF-1-specific coactivator of estrogen receptor. J Biol Chem 2001; 276:23962-23968.
154. Seroz T, Winkler GS, Auriol J et al. Cloning of a human homolog of the yeast nucleotide excision repair gene MMS19 and interaction with transcription repair factor TFIIH via the XPB and XPD helicases. Nucleic Acids Res 2000; 28:4506-4513.
155. Schwarz A, Stander S, Berneburg M et al. Interleukin-12 suppresses ultraviolet radiation-induced apoptosis by inducing DNA repair. Nat Cell Biol 2002:26-31.

156. Bergeron J, Benlimame N, Zeng N et al. Identification of the interferon-inducible double-stranded RNA-dependent protein kinase as a regulator of cellular response to bulky adducts. Cancer Res 2000; 60:6800-6804.
157. Suzuki Y, Sugita K, Suzuki N et al. Enhancement of XPG mRNA transcription by human interferon-beta in Cockayne syndrome cells with complementation group B. Int J Mol Med 1999; 3:87-89.
158. Gaspari AA, Fleisher TA, Kraemer KH. Impaired interferon production and natural killer cell activation in patients with the skin cancer-prone disorder, xeroderma pigmentosum. J Clin Invest 1993; 92:1135-1142.
159. Suzuki N, Suzuki H. Suppression of UV mutagenicity by human interferon. Mutat Res 1988; 202:179-183.
160. Hochstrasser M. Ubiquitin-dependent protein degradation. Ann Rev Genet 1996; 30:403-409.
161. Deshaies RJ. SCF and Cullin/Ring H2-based ubiquitin ligases. Annu Rev Cell Dev Biol 1999; 15:435-467.
162. Lommel L, Chen L, Madura K et al. The 26S proteasome negatively regulates the level of overall genomic nucleotide excision repair. Nucleic Acids Res 2000; 28:4839-4845.
163. Lommel L, Bucheli ME, Sweder KS. Transcription-coupled repair in yeast is independent from ubiquitylation of RNA pol II: Implications for Cockayne's syndrome. Proc Natl Acad Sci USA 2000; 97:9088-9092.
164. Schauber C, Chen L, Tongaonkar P et al. Rad23 links DNA repair to the ubiquitin/proteasome pathway. Nature 1998; 391:715-718.
165. Chen L, Shinde U, Ortolan TG et al. Ubiquitin-associated (UBA) domains in Rad23 bind ubiquitin and promote inhibition of multi-ubiquitin chain assembly. EMBO Reports 2001; 2:933-938.
166. Gillette TG, Huang W, Russell SJ et al. The 19S complex of the proteasome regulates nucleotide excision repair in yeast. Genes Dev 2001; 15:1528-1539.
167. Russell SJ, Reed SH, Huang W et al. The 19S regulatory complex of the proteasome functions independently of proteolysis in nucleotide excision repair. Mol Cell 1999; 3:687-695.
168. van Laar T, van der Eb AJ, Terleth C. A role for Rad23 proteins in 26S proteasome-dependent protein degradation? Mutat Res 2002; 499:53-61.
169. Chen Z, Zhang Y, Douglas L et al. UV-damaged DNA-binding proteins are targets of CUL-4A-mediated ubiquitination and degradation. J Biol Chem 2001; 276:48175-48182.
170. Svejstrup JQ. Mechanisms of transcription-coupled DNA repair. Nat Rev Mol Cell Biol 2002; 3:21-29.
171. Luo Z, Zheng J, Lu Y et al. Ultraviolet radiation alters the phosphorylation of RNA polymerase II large subunit and acceleration of its proteasome-dependent degradation. Mutat Res 2001; 486:259-274.
172. Weeda G, Rossignol M, Fraser RA et al. The XPB subunit of repair/transcription factor TFIIH directly interacts with SUG1, a subunit of the 26S proteasome and putative transcription factor. Nucleic Acids Res 1997; 25:2274-2283.
173. McKay BC, Chen F, Clark ST et al. UV-light-induced degradation of RNA polymerase II is dependent on the Cockayne's syndrome A or B protiens but not p53 or MHL1. Mutat Res 2001; 485:93-105.
174. Woudstra EC, Gilbert C, Fellows J et al. A Rad26-Def1 complex coordinates repair and RNA pol II proteolysis in response to DNA damage. Nature 2002; 415:929-933.
175. Gonzalez F, Delahodde A, Kodadek T et al. Recruitment of a 19S proteasome subcomplex to an activated promoter. Science 2002; 296:548-550.
176. Giaever G, Chu AM, Ni L et al. Functional profiling of the Saccharomyces cerevisiae genome. Nature 2002; 418:387-391.
177. Siede W, Robinson GW, Kalainov D et al. Regulation of the RAD2 gene of Saccharomyces cerevisiae. Mol Microbiol 1989; 3:1697-1707.
178. Cole GM, Mortimer RK. Failure to induce a DNA repair gene, RAD54, in Saccharomyces cerevisiae does not affect DNA repair or recombination phenotypes. Mol Cell Biol 1989; 9:3314-3322.

179. Hollander MC, Alamo I, Jackman J et al. Analysis of the mammalian gadd45 gene and its response to DNA damage. J Biol Chem 1993; 268:24385-24393.
180. Takahashi S, Saito S, Ohtani N et al. Involvement of the Oct-1 regulatoruy element of the gadd 45 promoter in the p53-independent response to ultraviolet damage. Cancer Res 2001; 61:1187-1195.
181. Latonen L, Taya Y, Laiho M. UV-radiation induces dose-dependent regulation of p53 response and modulates p53-HDM2 interaction in human fibroblasts. Oncogene 2001; 20:6784-6793.
182. Chen IT, Smith ML, O'Connor PM et al. Direct interaction of Gadd45 with PCNA and evidence for competitive interaction of Gadd45 and p21Wafl/Cip1 with PCNA. Oncogene 1995; 11:1931-1937.
183. Smith ML, Ford JM, Hollander MC et al. p53-mediated DNA repair responses to UV radiation: studies of mouse cells lacking p53, p21, and/or gadd45 genes. Mol Cell Biol 2000; 20:3705-3714.
184. Carr AM. Piecing together the p53 puzzle. Science 2000; 287:1765-1766.
185. Herrlich P, Ponta H, Rahmsdorf HJ. DNA damage-induced gene expression: signal transduction and relation to growth factor signalling. Rev Physiol Biochem Pharmacol 1992; 119:187-223.
186. Herrlich P, Rahmsdorf HJ. Transcriptional and posttranscriptional responses to DNA-damaging agents. Curr Opin Cell Biol 1994; 6:425-431.
187. Unsal-Kacmaz K, Makhov AM, Griffith JD et al. Preferential binding of ATR protein to UV-damaged DNA. Proc Natl Acad Sci USA 2002; 99:6673-6678.
188. Tibbetts RS, Cortez D, Brumbaugh KM et al. Functional interactions between BRCA1 and the checkpoint kinase ATR during genotoxic stress. Genes Dev 2000; 14:2989-3002.
189. Painter RB. Radioresistant DNA synthesis: An intrinsic feature of ataxia telangiectasia. Mutat Res 1981; 84:183-190.
190. Myung K, Kolodner RD. Suppression of genomic instability by redundant S-phase checkpoint pathways in Saccharomyces cerevisiae. Proc Natl Acad Sci USA 2002; 99:4500-4507.
191. Kaufmann WK, Wilson SJ. G1 arrest and cell-cycle-dependent clastogenesis in UV-irradiated human fibroblasts. Mutat Res 1994; 314:67-76.
192. Waga S, Hannon GJ, Beach D et al. The p21 inhibitor of cyclin-dependent kinases controls DNA replication by interaction with PCNA. Nature 1994; 369:574-578.
193. Kaufmann WK, Cleaver JE. Inhibition of replicon initiation and DNA chain elongation by ultraviolet light in normal human and xeroderma pigmentosum fibroblasts. J Mol Biol 1981; 149:171-187.
194. Hirao A, Kong Y-Y, Matsuoka S et al. DNA damage-induced activation of p53 by the checkpoint kinase Chk2. Science 2000; 287:1824-1827.
195. Lehmann AR, Kirk-Bell S, Arlett CF et al. Xeroderma pigmentosum cells with normal levels of excision repair have a defect on DNA synthesis after UV-irradiation. Proc Natl Acad Sci USA 1975; 72:219-235.
196. Park SD, Cleaver JE. Postreplication repair: Questions of its definition and possible alterations in xeroderma pigmentosum cell strains. Proc Natl Acad Sci USA 1979; 76:3927-3931.
197. Zernik-Kobak M, Vasunia K, Connelly M et al. Sites of UV-induced phosphorylation of the p34 subunit of replication protein A from HeLa cells. J Biol Chem 1997; 272:23896-23904.
198. Johnson RE, Kondratick CM, Prakash S et al. hRAD30 mutations in the variant form of xeroderma pigmentosum. Science 1999; 264:263-265.
199. Masutani C, Kusumoto R, Yamada A et al. The XPV (xeroderma pigmentosum variant) gene encodes human DNA polymerase η. Nature 1999; 399:700-704.
200. Ensch-Simon I, Burgers PM, Taylor JS. Bypass of a site-specific cis-syn thymine dimer in an SV40 vector during in vitro replication by HeLa and XPV cell-freee extracts. Biochemistry 1998; 37:8218-8226.
201. O'Day CL, Burgers PM, Taylor J-S. PCNA-induced DNA synthesis past cis-syn and trans-syn-I-thymine dimers by calf thymus DNA polymerase delta in vitro. Nucleic Acids Res 1992; 20:5403-5406.
202. Rauth AM. Evidence for dark reactivation of UV light damage in mouse L cells. Radiat Res 1967; 31:121-138.

203. Lehmann AR. The relationship between pyrimidine dimers and replicating DNA in ultraviolet light-irradiated human fibroblasts. Nucleic Acids Res 1979; 7:1901-1912.
204. Limoli CL, Laposa R, Cleaver JE. DNA Replication arrest in XP Variant cells after UV exposure is diverted into an Mre11-dependent recombination pathway by the kinase inhibitor Wortmannin. Mutat Res 2002; X:Y-Z.
205. Rowley R, Zhang J. Caffeine-mediated override of checkpoint controls. A requirement for rhp6 (Schizosaccharomyces pombe). Genetics 1999; 152:61-71.
206. Bailley V, Lauder S, Prakash S et al. Yeast DNA repair proteins Rad 6 and Rad 18 form a heterodimer that has ubiquitin conjugating, DNA binding and ATP hydrolytic activities. J Biol Chem 1997; 272:23360-23365.
207. Cong F, Hwang BJ, Vuong BQ et al. Interaction between UV-DDB protins and the c-Abl tyrosine kinase. J Biol Chem 2002.
208. Chen RH, Maher VM, McCormick JJ. Effect of excision repair by diploid human fibroblasts on the kinds and locations of mutations induced by (+/-)-7 beta,8 alpha-dihydroxy-9 alpha,10 alpha-epoxy-7,8,9,10-tetrahydrobenzo[a]pyrene in the coding region of the HPRT gene. Proc Natl Acad Sci USA 1990; 87:8680-8684.
209. Steitz TA. DNA polymerases: Structural diversity and common mechanisms. J Biol Chem 1999; 274:17395-17398.
210. Brash DE, Rudolph JA, Simon JA et al. A role for sunlight in skin cancer: UV-induced p53 mutations in squamous cell carcinoma. Proc Natl Acad Sci USA 1991; 88:10124-10128.
211. Ohmori H, Friedberg EC, Fuchs RPP et al. The Y-Family of DNA Polymerases. Mol Cell 2001; 8:7-8.
212. Cleaver JE. Stopping DNA replication in its tracks. Science 1999; 285:212-213.
213. Trincao J, Johnson RE, Escalante CR et al. Structure of the catalytic core of S. cerevisiae DNA polymerase η: Implications for translesion synthesis. Mol Cell 2001; 8:417-426.
214. Johnson RE, Washington MT, Prakash S et al. Fidelity of human DNA polymerase η. J Biol Chem 2000; 275:7447-7450.
215. Matsuda T, Bebenek K, Masutani C et al. Low fidelity DNA synthesis by human DNA polymerase eta. Nature 2000; 404:1011-1013.
216. Kannouche P, Broughton BC, Volker M et al. Domain structure, localization, and function of DNA polymerase eta, defective in xeroderma pigmentosum variant cells. Genes Dev 2001; 15:158-172.
217. Thakur M, Wernick M, Collins C et al. DNA polymerase h undergoes alternative splicing, protects against UV sensitivity and apoptosis, and suppresses Mre11-dependent recombination. Genes Chromosomes Cancer 2001; 32:222-235.
218. Haracska L, Kondratick CM, Unk I et al. Interaction with PCNA is essential for yeast DNA polymerase η function. Mol Cell 2001; 8:407-415.
219. Hoege C, Pfander B, Moldovan GL et al. RAD6-dependent DNA repair is linked to modification of PCNA by ubiquitin and SUMO. Nature 2002; 419:135-141.
220. Limoli CL, Giedzinski E, Morgan WF et al. Polymerase η deficiency in the XP variant uncovers an overlap between the S phase checkpoint and double strand break repair. Proc Natl Acad Sci USA 2000; 97:7939-7946.
221. Limoli CL, Giedzinski E, Bonner WM et al. UV-induced replication arrest in the xeroderma pigmentosum variant leads to double strand breaks, *γ-H2Ax formation, and Mre11 relocalization. Proc Natl Acad Sci USA 2002; 99:233-238.
222. Cleaver JE, Afzal V, Feeney L et al. Increased UV sensitivity and chromosomal instability related to p53 function in the xeroderma pigmentosum variant. Cancer Res 1999; 59:1102-1108.
223. Petrini JHJ. The mammalian Mre11-rad50-Nbs1 protein complex: integration of functions in the mammalian cellular DNA-damage response. Am J Hum Genet 1999; 64:1264-1269.
224. Nouspikel T, Hanawalt P. Terminally differentiated human neurons repair transcribed genes but display attenuated global DNA repair and modulation of repair gene expression. Mol Cell Biol 2000; 20:1562-1570.

225. Bill CA, Tofilon. Influence of DNA repair capacity and cell differentiation on UV-induced gene expression. Int J Rad Biol 1994; 65:299-306.
226. Ho L, Hanawalt PC. Gene-specific DNA repair in terminally differentiating rat myeloblasts. Mutat Res 1991; 255:123-141.
227. Islas AL, Hanawalt PC. DNA repair in the MYC and FMS proto-oncogenes in ultraviolet light-irradiated human HL-60 promyelocytes. Cancer Res 1995; 55:336-341.
228. Li G, Ho VC, Mitchell DL et al. Differentiation-dependent p53 regulation of nucleotide excision repair in keratinocytes. Am J Pathol 1997; 150:1457-1464.
229. Koberle B, Masters JR, Hartley JA et al. Defective repair of cisplatin-induced DNA damage caused by reduced XPA protein in testicular germ cell tumours. Curr Biol 1999; 9:273-276.
230. Einhorn LH. Curing metastatic testicular cancer. Proc Natl Acad Sci USA 2002; 99:4592-4595.
231. Mitchell DL, Volkmer B, Breitbart EW et al. Identification of a nondividing subpopulation of mouse and human epidermal cells exhibiting high levels of persistent UV photodamage. J Invest Dermatol 2001; 117:590-595.
232. Mohrenweiser HW, Jones IM. Variation in DNA repair is a factor in cancer susceptibility: A paradigm for the promises and perils of individual and population risk estimation? Mutat Res 1998; 400:15-24.
233. Qiao Y, Spitz MR, Shen H et al. Modulation of repair of ultraviolet damage in the host-cell reactivation assay by polymorphic XPC and XPD/ERCC2 genotypes. Carcinogenesis 2002; 23:295-299.
234. Spitz MR, Wu X, Wang Y et al. Modulation of nucleotide excision repair capacity by XPD polymorphisms in lung cancer patients. Cancer Res 2001:1354-1357.

CHAPTER 5

Mammalian DNA Mismatch Repair

Helen M.R. Robinson and Robert Brown

Abstract

Some DNA replication errors escape the proof reading activity of DNA polymerases and if allowed to persist will lead to mutations and potential creation of an abnormal mutant cell. Therefore such errors have to be repaired or the potentially abnormal cell needs to be eliminated. DNA mismatch repair (MMR) is one of the mechanisms by which cells achieve this by recognising and correcting mispairs and insertions occurring during DNA replication, as well as during recombination between diverged sequences. Loss of mismatch repair leads to a mutator phenotype and is associated with certain hereditary forms of cancer, such as Hereditary Non-Polyposis Colorectal Cancer and also many sporadic cancers. Furthermore, in response to certain types of DNA damage mismatch repair proteins are necessary for signalling biochemical pathways that lead to cell death by apoptosis. Thus, as well as a role in cancer susceptibility, loss of MMR is also associated with how tumour cells respond to many important anti-cancer agents. The following chapter describes the central MMR proteins, their role in MMR activity, their main interactions with other cellular proteins and discusses how MMR couples mismatch recognition to apoptotic signalling events.

What Is DNA Mismatch Repair

DNA mismatch repair (MMR) is the mechanism by which nucleotide mismatches are recognised, excised and then repaired by the MMR machinery.[1-4] These mismatches may be caused by misincorporation of an incorrect base by DNA Polymerases during replication, but also by physical damage to the DNA, such as caused by mono-functional alkylating agents.

During DNA replication, errors can occur at a rate of 10^{-5} or greater, the proof-reading activity of DNA polymerase α reduces this rate to less than 10^{-7} and then the mismatch repair process reduces this again to a rate less than 10^{-9}. Therefore, loss of MMR in cells leads to several hundred-fold increase in the mutation rate of genes. This occurs in a large proportion of patients suffering from the hereditary form of colorectal cancer Hereditary Non-polyposis Colorectal Cancer (HNPCC) and also in a proportion of patients with sporadic colorectal cancer and other tumor types, predominantly due to mutation or epigenetic silencing of the MLH1 and MSH2 MMR genes.[5-7] This mutator phenotype is readily identified in these cells by increased mutations at simple repeat sequences, microsatellites, and gives rise to the so-called microsatellite instability (MSI) phenotype.

MMR is also implicated in resistance to a variety of DNA damaging agents and loss of MLH1 has been reported to cause resistance to alkylating agents (e.g., temozolomide, MNU, MNNG), purine analogues (e.g., 6- thioguanine) and platinum coordination complexes (e.g., cisplatin and carboplatin). Many of these agents are clinically important anticancer drugs and

Eukaryotic DNA Damage Surveillance and Repair, edited by Keith W. Caldecott.

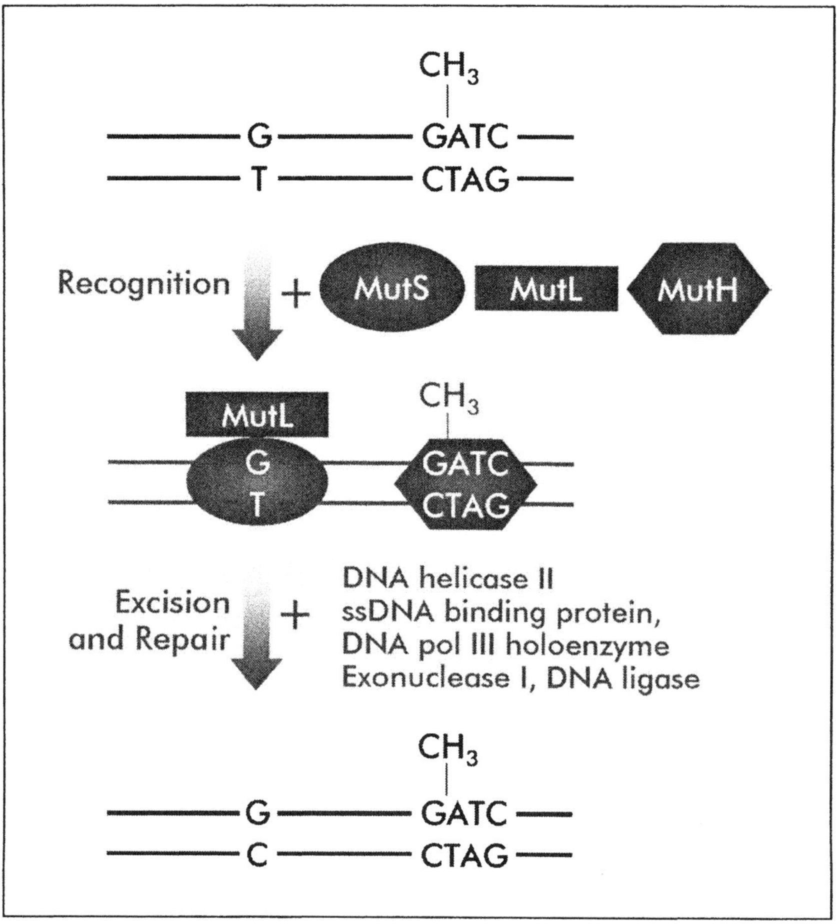

Figure 1 . Schematic of the prokaryotic mismatch repair mechanism. MutS binds to a mismatch, MutL then interacts with MutS and recruits the endonuclease MutH. A DNA excision and repair reaction then follows where repair is always directed to the newly synthesised strand.

studies suggest a role for MMR in cellular signalling pathways that lead to apoptosis and cell cycle arrest induced by these drugs.

The Mismatch Repair Mechanism

Although all the components of the eukaryotic mismatch correction system have not yet been identified, the relatively simple prokaryotic MMR process is well defined and has been reconstituted in vitro.[8] These reconstitution experiments showed that efficient MMR requires the 8 enzyme activities of MutL, MutS, MutH, DNA helicase II, ssDNA binding protein, DNA pol III holoenzyme, Exonuclease I and DNA ligase in the presence of ATP and dNTP's (Fig. 1).

During the MMR reaction MutS recognises and binds to the mismatch, which is reliant on the conformation change in DNA caused by the mispair. MutL interacts with MutS and activates the helicase MutH. MutH then translocates to an adjacent hemimethylated GATC site either 5' or 3' to the mismatch and creates a nick in the newly synthesised daughter DNA

Table 1. The MMR proteins and their homologues

E. coli Protein	*S. cervisiae* Protein	Human Protein
MutS	MSH2	MSH2
	MSH3	MSH3
	MSH6	MSH6 (GTBP/p160)
MutL	PMS1	PMS2
	MLH1	MLH1
	MLH2	PMS1
	MLH3	MLH3

strand. This step provides strand specificity to ensure that the newly synthesised, and not the parental strand, is repaired. Excision and resynthesis then occurs, where up to several kilobases of DNA spanning the mismatch may be removed and resynthesised.

The process of MMR in mammalian cells, although homologous, is more complex. Firstly, there are several MutL and MutS homologues (Table 1). The human MutL homologues are MLH1, PMS1, PMS2 and MLH3. The human MutS homlogues are MSH2, MSH3 and MSH6. These proteins act as heterodimers that potentially have slightly different roles in mismatch recognition and signalling.

MSH2 and MSH6 dimerise to form MutSα, whereas MSH2 and MSH3 dimerise to form MutSβ.[9,10] These heterodimers have slightly different specificities - MutSα primarily recognises G:T mismatches but both MutSα and MutSβ recognise short insertions / deletions.

MLH1 and PMS2 dimerise to form MutLα and defects in either component of the heterodimer abolish MMR.[11] MLH1 also forms heterodimers with PMS1 and MLH3. Although loss of MLH3 appears to lead to a mutator phenotype,[12] the significance of MLH1:PMS1 heterodimers for MMR is currently unclear.[13]

Although there is an abundance of MutL and MutS homologues identified in mammalian cells, there is so far a distinct lack of candidate MutH homologues - the enzyme responsible for strand discrimination in bacteria. Thus, the eukaryotic method of strand discrimination and the precise role of MLH1 remains unknown. Currently, the most widely accepted theory is that DNA replication intermediates are used as a strand discrimination signal, a notion supported by the fact that MLH1 and MSH2 can bind to PCNA, which is found at sites of DNA replication.[14-17]

There are currently three schools of thought as to how MMR proteins signal to proteins in the upstream excision and repair reaction. These are the sliding clamp model,[18-20] the ATP-dependent translocation model[21,22] and the induced fit model.[23,24]

In the sliding clamp model, the MutS complexes are proposed to act in a manner comparable with G-proteins, in that nucleotide exchange results in a conformational change, resulting in signal transduction to further biochemical cascades. In the ADP-bound form, MutSα has a high affinity for mismatched DNA. The mismatch then stimulates exchange of ADP for ATP, which results in a conformational change in the MutS heterodimer. The dimer is proposed to form a clamp around the DNA with a nonspecific affinity and freely slides down the DNA as a 'sliding clamp'. This then signals to other mismatch repair proteins and stimulates polymerase exonuclease activity, thus causing the excision and repair steps to proceed. ATP-hydrolysis then releases the MutS complex from the DNA, completing the MMR repair cycle.[18-20]

Similarly, in the ATP-dependent translocation model, following mismatch binding nucleotide exchange causes diffusion of the MutS complex away from the mismatch. However, in this model, instead of sliding down the DNA as a sliding clamp, the DNA is threaded through the middle of the complex using ATP hydrolysis to fuel translocation of the MutS heterodimer. This is supported by electron microscopy data where a mismatch-containing loop can be observed extruding from *E.coli* MutS homodimers.[21,22]

Thirdly, crystal structures of MutS bound to DNA mismatches have been elucidated.[23,24] These studies showed that MutS binding to mismatches causes a kink in the DNA. This led to the hypothesis that MutS mediates mismatch repair signalling directly through interactions with other MMR components i.e., interacting proteins are brought near MutS by bending the DNA. This 'induced fit' model would explain the observation that MutS can activate MutH activity on a separate DNA molecule in the presence of MutL and ATP.

The Mismatch Repair Proteins

MLH1

The human *mlh1* gene is situated on chromosome 3p21.3 and has an ORF of 2270bp that encodes for an 84.6 kDa, 756 amino acid long protein. The translated protein shows little sequence homology to any other proteins (except other MutL Homologs) and it's structural conformation provides few clues to its enzymatic activity.

Although the crystal structure of human MLH1 has not been elucidated, there has been considerable work on the structure of it's bacterial counterpart. The crystal structure of the highly conserved N-terminal 40kDa (residues 1-349) moiety of MutL (LN40) has been resolved and is shown to be a homodimeric ATPase that structurally resembles other ATPases such as DNA Gyrase and HSP90 (Fig. 2 and Fig. 3).[25]

This N-terminal MutL moiety (LN-40) was shown to contain two domains linked by two short helices to give an 'elbow' appearance. The first domain (residues 20-200) consists of one surface of eight stranded mixed β-sheet and another surface of five α-helices. The second domain is an α/β barrel with five mixed strands and two parallel helices. This second domain contains an unusual left-handed crossover of strand-helix-strand—a motif only observed in a small number of proteins including DNA Gyrase and the ribosomal translocase EF-G.[25]

Various biochemical tests were carried out on LN40. It was shown to weakly bind DNA in both a mismatch and sequence independent manner and shown to have ATP binding affinity. However, unlike intact MutL, LN40 does not have ATPase activity implying that either the MutL C-terminus or other proteins are required for MutL to function as an ATPase.[25] Intact MutL was shown to be a weak ATPase with a slow hydrolysis rate. This led to the hypothesis that MutL is a 'molecular timer' that uses it's slow ATP hydrolysis rate to allow time for complex assembly prior to the excision and repair reaction following mismatch recognition.

It was also deduced that addition of ATP or a nonhydrolysable ATP analogue causes a conformational change in the LN40 protein that allows interaction of the N-terminal domains of each protein in the homodimer. Furthermore, the ATPase activity was shown to require single-stranded DNA, thus MutL is a single-stranded DNA dependent ATPase.[25,26]

The N-terminus of the human MLH1 also acts as an ATPase. However, the C-terminus is responsible for interactions between the other MutL Homologs and other proteins that are described later in this chapter.

Experiments employing the yeast two hybrid technique and pulldown experiments have shown the interacting domain on MLH1 for binding to PMS1, PMS2 and MLH3 to be C-terminal[27,28] with the minimal binding domain likely to be somewhere between residues 492-742. It was also shown that a variety of MLH1 mutations in this region that occur in patients with HNPCC disrupt binding of MLH1 to PMS2 and cause loss of functional MMR in vitro.[27,29,30] Several of these mutants caused either deletion or substitution of a charged basic

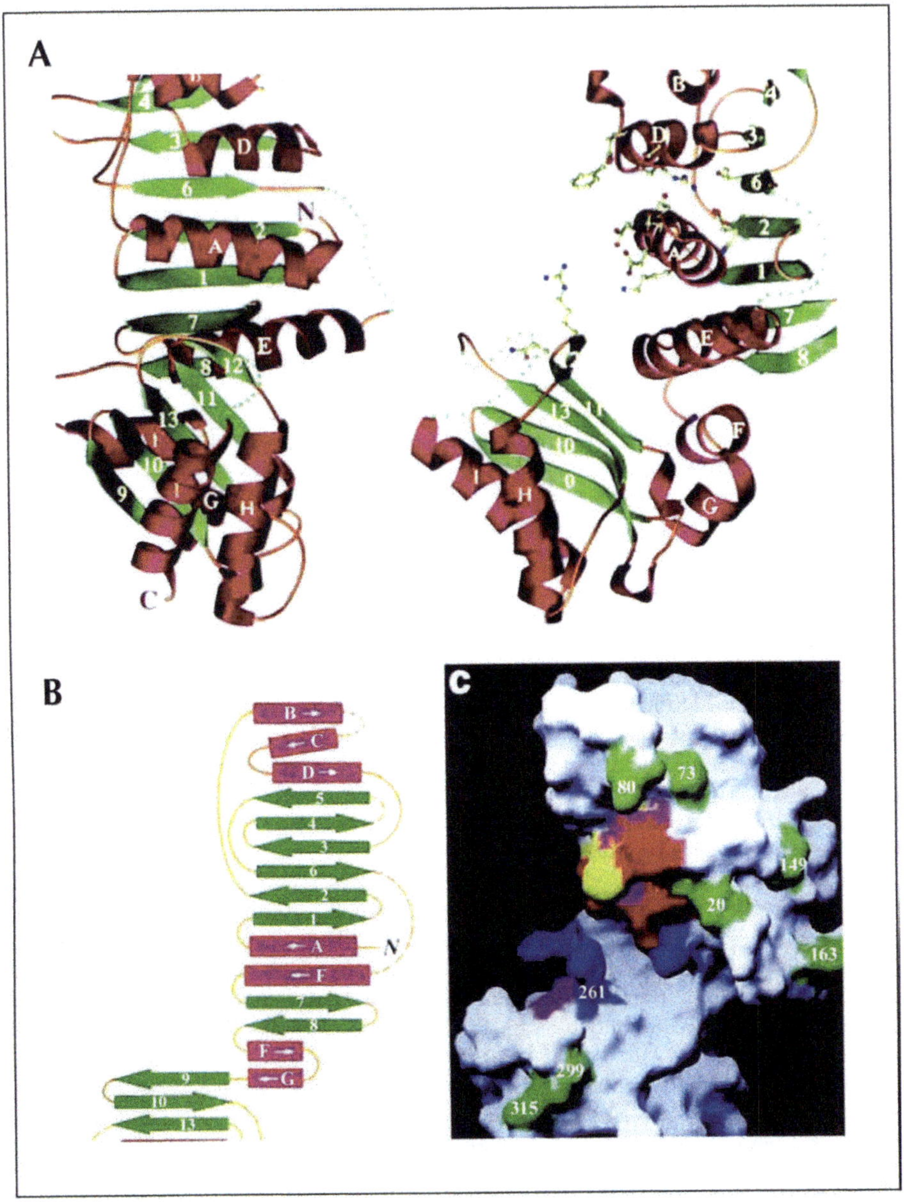

Figure 2. Crystal structure of the N-terminal 40kDa moiety of MutL. Reprinted with permission from Ban C, Yang W. Crystal structure and ATPase activity of MutL: implications for DNA repair and mutagenesis. Cell 1998; 95(4):541-52. © 2003 RG Landes Co. A) Orthogonal views of LN40 in ribbon diagram. Strands are shown as arrows, 310 helices as red coils, and loops as yellow lines. In one of the views, conserved residues in the putative ATP-binding sites are shown in ball-and-stick. The disordered, yet conserved, loops are illustrated as green dotted lines. B) Topology of LN40. Secondary structures are labelled. Four internal disordered loops are shown as dashed lines. C) Molecular surface of LN40 shows the two solvent-exposed patches of conserved residues. Conserved hydrophobic residues are shaded in yellow, basic in blue, acidic in red, and others in purple. The green patches represent the ends of disordered yet conserved loops.

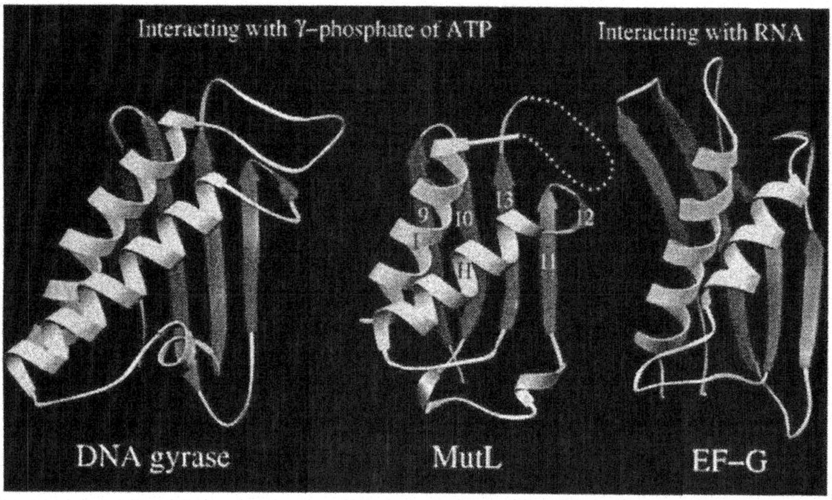

Figure 3. Ribbon diagram of a structurally similar domain in MutL, DNA Gyrase, and EF-G. Reprinted with permission from Ban C, Yang W. Crystal structure and ATPase activity of MutL: implications for DNA repair and mutagenesis. Cell 1998;95(4):541-52. © 2003 RG Landes Co.

lysine residue to the chemically unreactive, hydrophobic alanine or to the uncharged threonine. This implies that there is a charge based interaction between MLH1 and PMS2.

PMS2, PMS1 and MLH3

The human *pms2* gene is situated at chromosome 7p22 and encodes an 862 residue, 95.8 kDa protein. PMS2 is a homolog of MLH1 with which it forms the heterodimer MutLα in both yeast and humans.[27,28,31] Like MLH1, its N-terminal (residues 1-365) crystal structure has been resolved and it has been shown to be a weak ATPase.[32]

Experiments involving this 40kDa fragment showed that it has intrinsic DNA binding ability with a preference for single-stranded DNA. Incubation with radiolabelled ATP also showed that this fragment has ATPase activity that is unaffected by the presence of DNA but is inhibited by nonhydrolysable ATP analogs and mutations in the ATPase domain. Unexpectedly, this fragment was able to act as a monomer—a characteristic that is absent from its structural homologues.

The structure of PMS2 closely resembles LN40 and, like the MutL fragment, the PMS2 fragment consists of two domains. The first domain (residues 29-221) consists of eight strands of mixed β-sheet packed against a layer of five α-helices. Two α-helices connect this domain to the second domain (residues 239-365) that has an α/β barrel structure consisting of five strands of β-mixed sheet and 3 α-helices with a further two short β-strands covering on the end of the barrel.

Also like MLH1, the C-terminus of PMS2 is where it interacts with other MutL homologues. Unlike MLH1 however, this is a small region located between residues 612-674 as defined by yeast two hybrid and pulldown experiments.[28] This region was subjected to further analysis *in silico* and was shown to be a putative coiled-coil domain.

MLH3 also has a putative coiled-coil interaction domain (residues 860-895) that was identified based on its homology to the PMS2 motif. A homologous region on PMS1 (residues 693-728) was not predicted to be a coiled-coil. These regions were tested for binding to MLH1

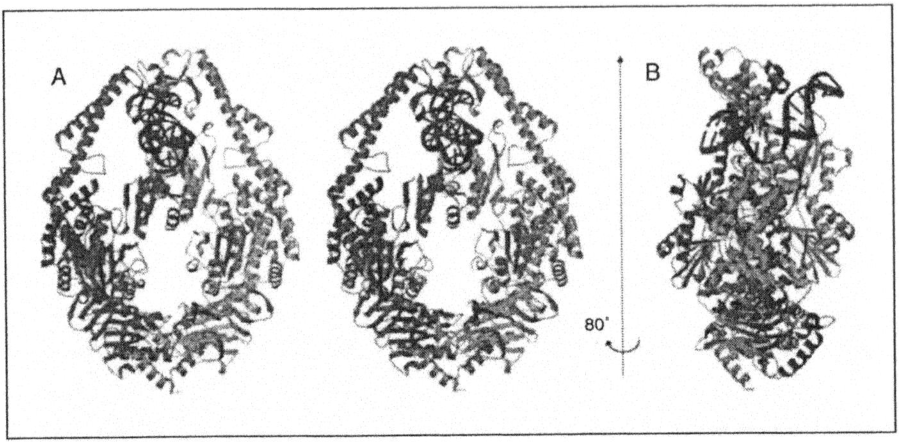

Figure 4. The MutS:DNA complex. Reprinted with permission from Lamers MH, Perrakis A, Enzlin JH, et al. The crystal structure of DNA mismatch repair protein MutS binding to a G x T mismatch. Nature 2000;407(6805):711-7. © 2003 RG Landes Co.

using yeast two hybrid experiments and were shown to strongly interact. Furthermore, it was shown using pulldown experiments that high levels of MLH3 and PMS1 can compete with PMS2 for binding to MLH1.[28] This suggests that the balance of expression of MMR proteins may be critical for function. However there is little available information about the relative expression of the MutL homologues in cells.

The MutS Homolog Proteins

MSH2

The *msh2* gene is located on chromosome 2 at the p22-p21 locus. The gene encodes a 934 residue protein with a mass of 104.7 kDa. The crystal structure of MutS bound to a DNA mismatched substrate has been resolved[23,24] and shows this asymmetric MutS dimer to resemble a pair of 'praying hands' where only one monomer specifically binds to the mismatch (Fig. 4).

The N-terminal 'thumbs' (res 2-115) form the mismatch recognition domain, followed by a connector domain and then a core domain that are described as the 'palm of the hand'. The core domain (res 267-433 and 504-567) forms a helical bundle from which two helical 'levers' extend and embrace but do not touch the DNA. The clamp domain or 'fingertips' (res 444-503) now follow. The ATPase domain is attached to the core domain and shows a classical Walker A motif and there is also a helix-turn-helix motif involved in dimer contacts (res 766-800) (Fig. 5).[24]

Although homodimeric, as only one monomer binds to the mismatch, the dimer acts as a structural heterodimer. In humans MSH3 and MSH6 confer specificity of mismatch recognition and thus it is highly likely that MSH2 is the unbound monomer and MSH3/6 is the bound monomer. The bound monomer was shown to be ADP bound implying an increased affinity for DNA in the ADP bound state.[24]

MSH2 was tested for binding to MSH3 and MSH6 by pulldown experiments and shown to bind to each of these proteins. The interacting region was mapped and shown to consist of two domains flanking the ATPase domain at residues 378-625 and 875-934.[33] As the interac-

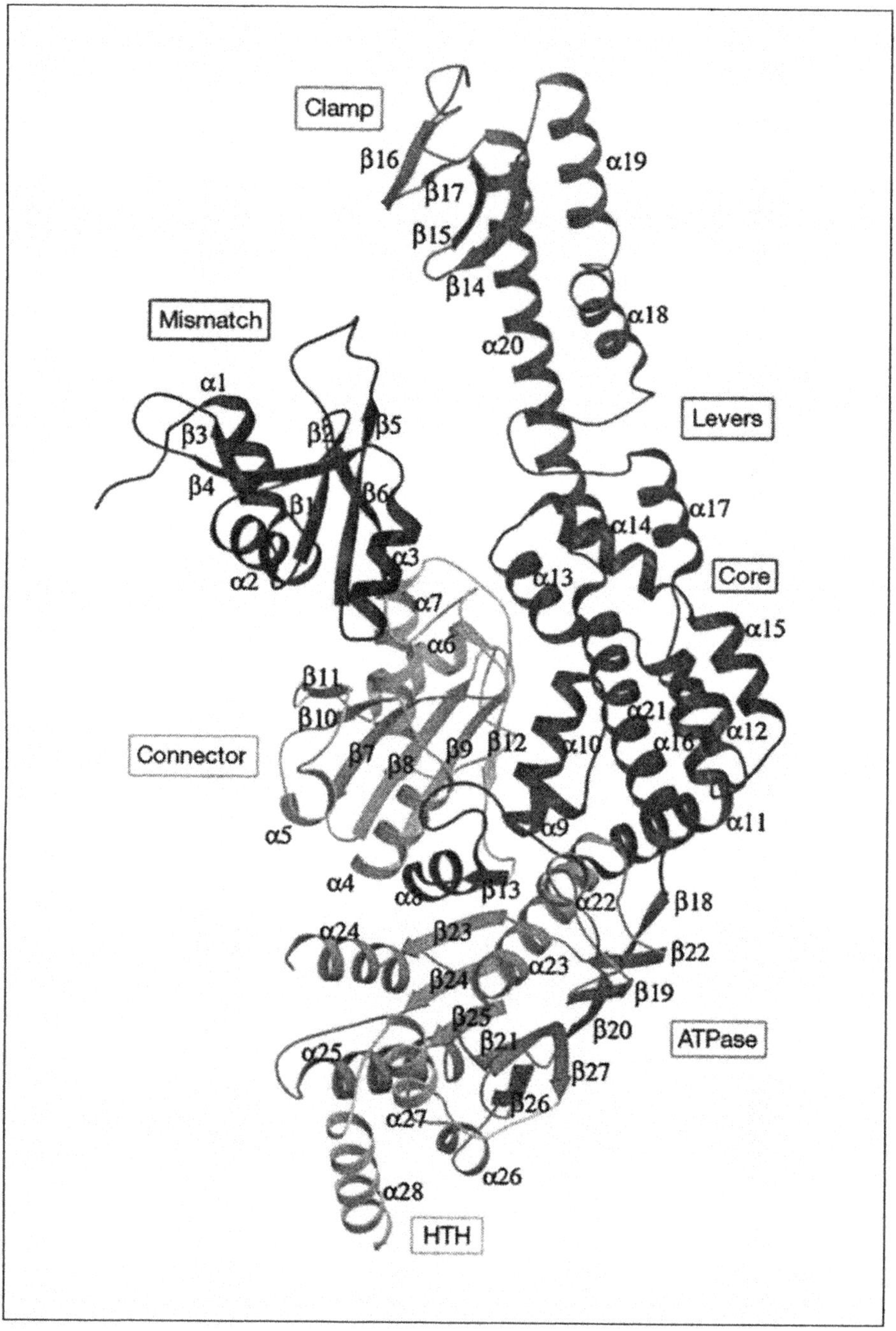

Figure 5. The mismatch binding monomer of the MutS homodimer. Reprinted with permission from Lamers MH, Perrakis A, Enzlin JH, et al. The crystal structure of DNA mismatch repair protein MutS binding to a G x T mismatch. Nature 2000;407(6805):711-7. © 2003 RG Landes Co.

tion regions flank the nucleotide binding regions it is hypothesised that interactions are regulated by nucleotide binding. Similar experiments were then performed with MSH2 with point mutants in the interacting domains that occur in patients with HNPCC.[33] Surprisingly these had no effect on dimer formation implying further functions of these regions.

However, another study discovered that various HNPCC missense mutations throughout MSH2 affect mismatch binding, ATP hydrolysis, ATP binding, nucleotide exchange and ATP dependent dissociation of MutSα from a mismatch.[34] Overall the results showed that HNPCC mutations impair the ability of MutSα to load multiple ATP bound complexes onto DNA.[34]

Recently it has been discovered that MSH2 is subject to post translational modifications.[35] There are several putative phosphorylation sites on MSH2 and in vitro phosphorylation experiments showed phosphorylation of both MSH2 and MSH6 by PKC and CKII. Furthermore, in vivo experiments showed phosphorylation of both these proteins, with MSH6 being most efficiently phosphorylated. A higher proportion of phosphorylated MutSα is observed in the nucleus than in the cytosol and dephosphorylation confers loss of binding to a G:T mismatch in an EMSA.[35] This is the first time that a MMR protein has been shown to be post translationally modified. It will be interesting to discover which residues are phosphorylated and whether these are important for HNPCC development.

MSH3 and MSH6

The human *msh6* gene is located at chromosome 2p16 and encodes a 1360 residue protein of 152.9 kDa that with MSH2 forms MutSα. The human *msh3* gene is located at chromosome 5q11-q12 and encodes a 1137 residue, 127.4 kDa protein that with MSH2 forms the heterodimer MutSβ.

Unlike MSH2, there is no structural information for MSH3 or MSH6 but both proteins are homologous to MSH2 in that they each have an ATPase domain flanked by two protein interaction regions.[33]

With MSH2, MSH6 forms MutSα via its interacting domains at residues 326-575 and 1302-1360. MSH3 interacts with MSH2 through residues 126-250 and 1050-1128. Like MSH2 it has been shown that MSH6 is phosophorylated post-translationally and this modification appears to regulate mismatch binding.[35] In fact, MSH6 was shown to be more efficiently phosphorylated than MSH2. To date, no other proteins in the MMR process have been shown to be subject to post translational modifications.

Proteins That Interact with the MuL and MutS Homologues

As well as interacting with their homologues, mismatch repair proteins especially MLH1 and MSH2 have other binding partners (Fig. 6). These interactions can give clues to how the MMR mechanism operates. Some of the most well characterised interactions are therefore discussed below.

Exonuclease 1

Exonuclease 1 (EXO1) is a 5'-3' double stranded DNA exonuclease that has been shown to interact with yeast MLH1[36] and is required for MMR repair in *E.Coli.*[37]

Human Exonuclease 1 (hEXO1) was identified as a human MLH1 interactor by yeast two hybrid screen[38] and was then also shown to interact with MLH1 in other systems i.e coimmunoprecipition experiments and pulldown experiments.[38,39] Similarly, hEXO1 has been shown to interact with MSH2 and MSH3,[39,40] although the interaction between MLH1 and hEXO1 is seen to be the most robust.[39]

Analysis of these interactions to determine the important regions on hEXO1 for protein binding show that residues 388-490 and 787-846 are essential for MLH1 binding. Residues

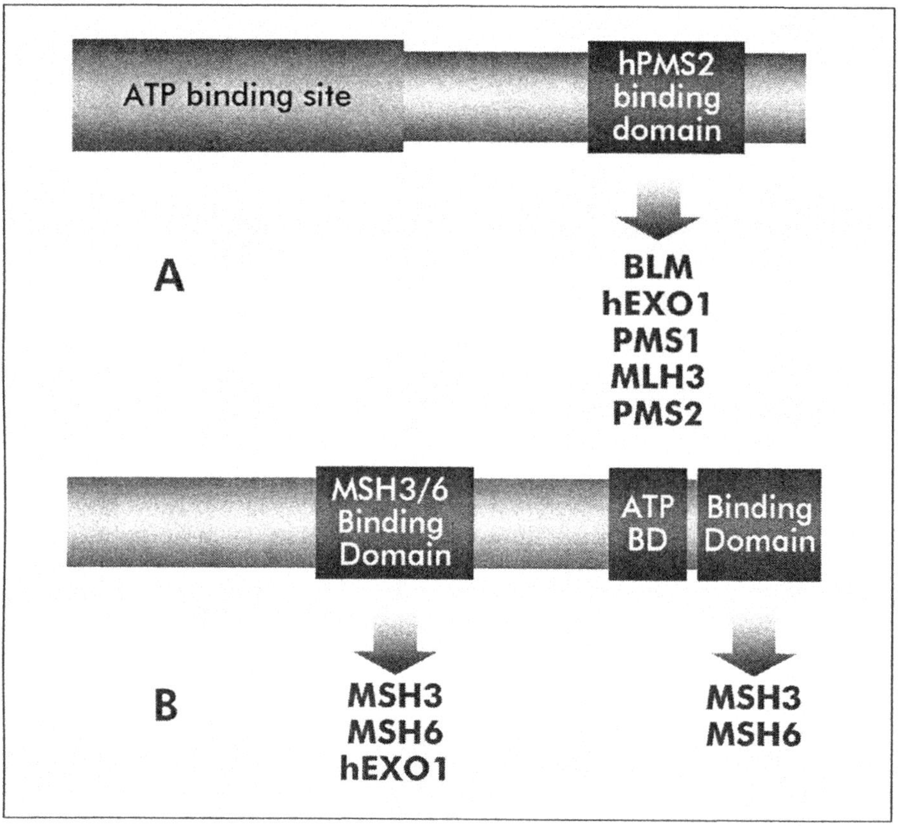

Figure 6. MLH1 and MSH2 bind to a variety of proteins.

604-846 are required for MSH2 binding, and residues 129-387 are needed for interaction with MSH3. This analysis was also carried out reciprocally, showing the binding site for hEXO1 on MSH2 to be residues 601-671, further stabilised by residues 261-600. Important residues on MSH3 are 75-297 and for MLH1 residues 410-650 are required for hEXO1 binding. This coincides with the binding site on MLH1 for PMS2 but interestingly hEXO1 does not compete with PMS2 for binding to MLH1.[39]

Further experiments used MLH1 proteins that contained mutations that occur in patients with HNPCC. Many of the mutants tested displayed reduction or complete loss of binding to hEXO1 both in the yeast two hybrid system and using pulldowns with in vitro transcribed and translated proteins.[38,39]

Furthermore, there is now evidence that hEXO1 is mutated in a subset of HNPCC patients and is also involved in both 5' and 3' human MMR in vitro.[41,42] As well as being mutated in HNPCC patients, hEXO1 is overexpressed in several liver cancer cell lines and also in colon and pancreas adenocarcinoma tumors but not in the normal controls for the corresponding tissues.[40] These studies, and those outlined above, especially those showing disruption of the interaction between hEXO1 and MMR proteins by HNPCC mutations, provide physiological significance. However, further biochemical studies need to be performed to explore the precise function of this interaction and the role of hEXO1 in DNA repair and cancer.

MED1 / MBD4

Methyl-CpG Binding Endonuclease I (MED1) was discovered in 1999 as an interactor of MLH1 in a yeast two hybrid screen[43] and also in 1998 by a different group as a homologue of MeCP1 who termed it MBD4.[44] The full length cDNA was shown to have an ORF of 1.7kb encoding a protein of 580 amino acids. The predicted structure of MED1 has two main domains, an amino-terminal methyl binding domain and a carboxy-terminal endonuclease domain connected by a putative nuclear localisation signal. The methyl binding domain (MBD) is homologous to other methyl CpG binding proteins such as MeCP2 and PCM1/MBD1 and the endonuclease domain is homologous to various damage specific endonucleases with glycosylase / lyase activity including MutY.

Electrophoretic Mobility Shift Assays (EMSAs) with methylated, hemimethylated and unmethylated oligonucleotides showed that MED1 can bind to hemi- and fully methylated DNA but with a higher affinity for fully methylated oligonucleotides. It was also shown that MED1 has a nonspecific endonuclease activity.[43]

Due to the methyl binding ability of MED1, it's ability to interact with MLH1 and early results implying a role for MED1 in MMR, it was initially hypothesised that MED1 was operating in MMR as a strand discrimination signal analogous to MutH.[43] However, further experiments using MMR assays utilizing a C:A substrate that was either unmethylated, hemi-methylated or fully methylated showed that methylation was unlikely to be the strand discrimination signal in humans.[45] In fact it is more likely that MED1 is involved in base excision repair (BER), not MMR, and the interaction between MED1 and MLH1 is an example of cross-talk between these different DNA repair pathways.

Proliferating Cell Nuclear Antigen

The Proliferating Cell Nuclear Antigen (PCNA) heterodimer is the processivity component of DNA polymerases δ and ε and functions as a 'sliding clamp' around the DNA. In yeast, PCNA interacts with the MSH2:MSH3 complex[16] and also the MSH2:MSH6 complex.[14] In humans, PCNA interacts with both MLH1 and MSH2 in an ATP-dependent fashion as seen by coimmunoprecipitation experiments.[15,17]

Yeast and human MSH6 and MSH3 but not MSH2 have a consensus PCNA binding motif in their amino-terminus of QXX(I/L/M)XX(F)(F/Y).[46] If this site is mutated then the interaction between these MutS homologues and PCNA is lost and the mutated MSH6 cannot fully complement an *msh3 msh6* yeast strain where wild type MSH6 can.[14]

EMSAs have shown that at low concentrations of MSH2:MSH6, PCNA causes an increase in the formation of a specific MMR complex on an oligonucleotide containing a G:T or G:C mismatch. At high MSH2:MSH6 levels, PCNA causes an increase in the formation of a specific complex as well as a decrease in the formation of a nonspecific complex. Addition of ATP or a nonhydrolysable ATP analogue ATPγS causes dissociation of this complex and inhibits MSH2:MSH6:PCNA binding.[14]

Further genetic studies in yeast have shown that PCNA is essential for MMR using yeast strains with mutations in the gene that encodes for PCNA - *pol30*. *Pol30-104* is a mutation that causes a C-T transition at nucleotide 752 resulting in an alanine to valine missense mutation at residue 251. This causes an 80 fold increase in genetic instability in simple repetitive sequences. Double mutants with *pol30-104* and *mlh1delta* or *msh2delta* are not synergistic, therefore implying that both gene products operate in the same system.[16]

In humans, PCNA has been shown to be required for MMR in vitro.[17,15] Cell extracts that were depleted of PCNA lost MMR activity that was then restored upon reintroduction of PCNA. Also p21 and a p21 peptide, both of which bind to PCNA, inhibit MMR when added to MMR proficient cell extracts. This can again be restored by increasing the concentration of PCNA in the extracts.[15]

Although it is clear that PCNA is required for MMR function, the precise role for PCNA in MMR remains obscure. There are several possible roles that can be envisioned. PCNA could act as a 'molecular matchmaker' through its interactions with several MutS and MutL homologues, stimulating complex formation between these MMR components. It could also act to load the MMR machinery onto DNA, or allow the MMR components to travel along the DNA via its interactions with both the DNA replication machinery and the MutS and MutL homologues. Another interesting possibility is that PCNA acts as the strand discrimination signal in MMR as it resides at DNA primer termini, a notion supported by the fact that PCNA and hMSH6 or hMSH3 colocalise to replication foci.[47]

The Bloom's Syndrome Protein and the BASC Complex

BLM is the gene mutated in Bloom's Syndrome (BS), a rare genetic disorder that is characterised by a variety of features including genomic instability and cancer predisposition. BLM is a 1417 amino acid protein that is a member of the RecQ family of DNA helicases.[48]

MLH1 was identified as a BLM interactor when it was pulled out in a yeast two hybrid screen using BLM as bait.[49,50] It was shown that MLH1 and MutLα but not PMS2 can interact with BLM and further analysis showed the essential residues on BLM for binding are 1-131, 448-572 and 1034-1417. The binding domain for BLM on MLH1 is in the carboxy-terminus between residues 396-756.

Although BLM interacts with MLH1, it appears that BLM is not important in MMR as BS cell lines are MMR proficient.[49,50] Interestingly, both BLM and MLH1 reside in the BRCA1 Associated Genome Surveillance Complex (BASC).[51] This is a large complex that contains a wide variety of tumor suppressors and oncogenes that function in DNA repair including BLM, MLH1, MSH2, MSH6, ATM, RFC, the RAD50-MRE11-NBS1 complex and also BRCA1. Staining of this complex by indirect immunofluorescence showed greater assembly of BASC after addition of hydroxyurea and in mid-late S phase or G2 phase of the cell cycle. This led to the hypothesis that the complex functions to coordinate DNA repair and the maintenance of genome integrity either as a sensor of DNA damage or by regulating post-replicational repair.[51]

Apoptosis and Drug Resistance

As well as causing a mutator phenotype, loss of mismatch repair can cause drug resistance to a variety of DNA damaging agents including some important anti-cancer drugs e.g., alkylating agents and platinum coordination complexes. This is proposed to be due to loss of a MMR dependent apoptotic signalling pathway that is utilised by these chemotherapeutic drugs to induce cell death. When such pathways are lost, DNA damage becomes tolerated and the cells do not undergo apoptosis.

There are three main models for how apoptotic signals are created via the MMR apparatus. The first of these is the 'Futile Repair Cycle' model.[52] This model suggests that during DNA replication the base that forms the most stable hydrogen-bonding is incorporated, but in the presence of certain types of base damage this need not be the complementary base. For example, O^6-methylguanine is formed in DNA by mono-functional methylating agents (e.g., temozolomide, MNNG, MNU etc) and is believed to be the main cytotic lesion induced by these agents. The base that pairs best with O^6-methylguanine is thymine and is thus incorporated into the new strand during replication forming a G:T base-pair rather than the normal G:C. However, the MMR apparatus recognises the G:T as a mismatch and attempts to repair it, but DNA synthesis during the resythesis step of MMR again incorporates thymine. Thus it is the mismatch on the newly synthesised strand that is removed rather than the O^6-methylguanine. It is therefore proposed that this results in repeated, futile cycles of repair that could possibly serve as a signal for apoptosis. If MMR were abolished there would be no attempt to repair the DNA damage, no futile cycles and thus damage is tolerated causing drug resistance.

A second model of apoptotic signalling via MMR is the 'MMR Dependent Replication Stalling' model.[53] Under physiological conditions, cisplatin is converted into a charged electrophilic molecule which forms cross-links with nucleophilic sites in the DNA. These cross-links usually take the form of 1,2-intrastrand cross-links between adjacent purines e.g., 1,2-d(GpG). This type of adduct is a poor substrate for hMutSα but after a round of replication, structures form that are bound much more tightly by MutSα, namely adducts where the platinated guanines are opposite noncomplementary bases.[54,55] The tight binding of MutSα to these adducts may be part of a mechanism that acts by stalling the replication machinery and causing a signal for apoptosis to be generated. If MMR were lost, then replication would not be stalled and this signal could not be generated leading DNA damage tolerance and drug resistance. Evidence for this model comes from Vaisman et al[56] who found using steady-state elongation assays that hMLH1 and hMSH6 defects cause a 1.5-4.8 fold increased cisplatin resistance and a 2.5-6 fold increase in replicative bypass of cisplatin adducts.

The third model is that of modulation by 'recombination bypass'.[57] It has been shown that RAD52, a protein required for recombination, mediates resistance in MMR defective yeast. This leads to a hypothesis that increased recombination or strand-exchange during DNA replication could bypass un-repaired damage thus causing damage tolerance and drug resistance. MMR could conceivably inhibit RAD52 mediated recombination bypass of DNA damage, an idea that is supported by the evidence that there is no drug resistance in MMR and *rad52* double mutants. Loss of MMR proteins would then cause an increase in bypass DNA lesions during replication and therefore reduced replication stalling, leading to damage tolerance and drug resistance.

Although it is clear that mismatch repair status is critical for cells to respond to certain types of DNA damage, it remains unclear precisely how MMR may engage or signal an apoptotic pathway. Thus it is unclear from the above models whether MMR is generating a lesion during DNA replication that is recognised by cell death pathways (for instance a DNA double strand break) or whether MMR is directly involved in signalling to apoptosis pathways. Studies in transgenic mice have shown that although complete loss of MSH2 expression is required to produce a mutator phenotype and loss of MMR activity, effects on resistance to alkylating agents are seen at various MSH2 expression levels.[58] This argues that absence of MMR per se is not required for resistance and favours a role for MSH2 in modulating or signalling apoptosis.

MMR and Apoptotic Pathways

Although it is clear that a MMR dependent apoptotic signal is generated following DNA damage, it is unclear which molecules are involved in this pathway. Interactions of proteins involved in apoptosis with MLH1 and MSH2 have been observed, including ATM and CHKII.[59] Loss of MMR has also been shown to influence the response of a variety of stress response pathways to DNA damage, including p73, c-abl, JNK and p53,[60-65] although the critical pathways may depend on the cell types examined. There has been particular focus on the role of p53 in transducing DNA damage signals due to its well defined role in transduction of apoptotic signals, particularly from DNA damage such as double strand breaks.

It has been shown in the murine intestine that loss of p53 causes loss of an immediate apoptotic response to temozolomide and cisplatin, but retention of a delayed response implying a partial p53 dependency on apoptosis. However, in cells where both p53 and MSH2 are lost, then the delayed p53 independent response is abolished, showing that there are both p53 dependent and MMR dependent, p53 independent mechanisms in place.[64]

Following treatment with alkylating agents, p53 is phosphorylated (and therefore activated) in a MutSα and MutLα dependent fashion.[61] However, although p53 becomes phosphorylated, this is not necessarily indicative of an apoptotic response. Indeed, p53 also has a major role in cell checkpoints and ser-15 and/or ser-392 phosphorylation may suggest checkpoint signalling, where ser-46 phosphorylation suggests apoptotic signalling.

It has also been shown that in response to ionising radiation loss of p53 and loss of PMS2 causes increased resistance to apoptosis in cell lines derived from knock out mice.[65] However, in the double knockouts, the effects of losing PMS2 and p53 were additive. This is convincing evidence that the MMR- dependent response to DNA damage is p53-independent.

On the whole it would appear that the MMR-dependent response to DNA damage is largely p53 independent, although this almost certainly will depend on the cell type and also the specific DNA damaging agent used.

If the MMR dependent apoptotic response occurs independently of p53, which proteins are required for signalling? The non-receptor tyrosine kinase c-Abl and also the c-Jun NH2 terminal kinase (JNK) have been shown to be induced in response to DNA damaging agents including ionising radiation, mitomycin C and cisplatin.[60,62,63] In cells that are mutant for MLH1 and MSH2 c-Abl activation was reduced in response to cisplatin and was then restored upon transfection of chromosome 3 into the MLH1 defective cell line and chromosome 2 into the MSH2 defective cell line.

Further to this it was shown using mutant mouse embryo fibroblasts (MEFs) that the p53 related protein p73 is stabilised in response to cisplatin in a c-Abl and MLH1 dependent manner. Consistent with this a neuroblastoma cell line that is defective for p73 but not p53 displayed increased resistance to cisplatin induced apoptosis. Furthermore, coexpression of c-Abl and p73 causes cell death in 3T3 cells and c-Abl null MEFs are resistant to killing by cisplatin.[60]

It would appear then that p73 and c-Abl could be the main players in MMR dependent apoptosis in response to cisplatin. However there is no available data on p73 response to other DNA damaging agents that cause damage thought to be detected by the MMR system. Also it is as yet unclear what the upstream signalling proteins are. JNK1 is a candidate, although activation of this protein in response to cisplatin does not appear to be universal.[63]

It has been shown that MLH1 interacts with the ataxia-telangiectasia protein ATM and that MSH2 interacts with CHKII both in vitro and in vivo.[59] Furthermore, in cells deficient in MMR, ATM-dependent phosphorylation of CHKII and down regulation of CDC25A was abolished, leading to aberrant DNA synthesis following irradiation. This implies a role for MMR proteins as adaptors that directly link DNA damage to downstream effector molecules. In this case, the experiments revealed MMR proteins functioning in this way to activate an S-phase checkpoint. Perhaps further work will produce evidence of MMR proteins similarly activating an apoptotic response to DNA damage?

Conclusions

The biochemical processes of eukaryotic DNA MMR are beginning to be unravelled. Detailed molecular characterisation of MutS homologues now provides models of how these proteins recognise mismatches and damage in DNA. The role of MutL homologues, although clearly critical, remains obscure. How MMR couples to DNA replication and how strand selectivity are achieved also remain unresolved, however the interaction of MMR proteins with cellular proteins such as PCNA begins to give clues how this may be achieved.

Research on MMR so far describes a highly important cellular process that is relevant to both the generation and treatment of a variety of both sporadic and hereditary cancers. While there has been major headway in unravelling the complexity of eukaryotic MMR mechanism there are still many questions to be answered. What is the strand discrimiation signal that allows the correct mispaired base to be removed? How are MMR-dependent apoptotic and cell cycle checkpoint signals generated induced by certain anti-cancer drugs? If these questions can be answered we will surely gain significant insight into carcinogenesis and chemotherapeutic drug resistance.

References

1. Bellacosa A. Functional interactions and signaling properties of mammalian DNA mismatch repair proteins. Cell Death Differ 2001; 8(11):1076-92.
2. Buermeyer AB, Deschenes SM, Baker SM et al. Mammalian DNA mismatch repair. Annu Rev Genet 1999; 33:533-64.
3. Kolodner R. Biochemistry and genetics of eukaryotic mismatch repair. Genes Dev 1996; 10:1433-1442.
4. Kolodner RD, Marsischky GT. Eukaryotic DNA mismatch repair. Curr Opin Gnet Dev 1999; 9:89-96.
5. Kinzler KW, Vogelstein B. Lessons from hereditary colorectal cancer. Cell 1996; 87(2):159-70.
6. Toft NJ, Arends MJ. DNA mismatch repair and colorectal cancer. J Pathol 1998; 185(2):123-9.
7. Jacob S, Praz F. DNA mismatch repair defects: Role in colorectal carcinogenesis. Biochimie 2002; 84(1):27-47.
8. Lahue RS, Au KG, Modrich P. DNA mismatch correction in a defined system. Science 1989; 245(4914):160-4.
9. Drummond JT, Li G-M, Longley MJ et al. Isolation of a hMSH2-p160 heterodimer that restores DNA mismatch repair in tumour cells. Science 1995; 268:1909-1912.
10. Genschel J, Littman SJ, Drummond JT et al. Isolation of MutSbeta from human cells and comparison of the mismatch repair specificities of MutSbeta and MutSalpha [published erratum appears in J Biol Chem 1998 Oct 9; 273(41):27034]. J Biol Chem 1998; 273(31):19895-901.
11. Li GM, Modrich P. Restoration of mismatch repair to nuclear extracts of H6 colorectal tumor cells by a heterodimer of human MutL homologs. Proc Natl Acad Sci USA 1995; 92(6):1950-4.
12. Lipkin SM, Wang V, Jacoby R et al. MLH3: A DNA mismatch repair gene associated with mammalian microsatellite instability. Nat Genet 2000; 24(1):27-35.
13. Raschle M, Marra G, Nystrom-Lahti M et al. Identification of hMutLbeta, a heterodimer of hMLH1 and hPMS1. J Biol Chem 1999; 274(45):32368-75.
14. Flores-Rozas H, Clark D, Kolodner RD. Proliferating cell nuclear antigen and Msh2p-Msh6p interact to form an active mispair recognition complex. Nat Genet 2000; 26(3):375-8.
15. Gu L, Hong Y, McCulloch S et al. ATP-dependent interaction of human mismatch repair proteins and dual role of PCNA in mismatch repair. Nucleic Acids Res 1998; 26(5):1173-8.
16. Johnson RE, Kovvali GK, Guzder SN et al. Evidence for involvement of yeast proliferating cell nuclear antigen in DNA mismatch repair. J Biol Chem 1996; 271(45):27987-90.
17. Umar A, Buermeyer AB, Simon JA et al. Requirement for PCNA in DNA mismatch repair at a step preceding DNA resynthesis. Cell 1996; 87(1):65-73.
18. Gradia S, Subramanian D, Wilson T et al. hMSH2-hMSH6 forms a hydrolysis-independent sliding clamp on mismatched DNA. Mol Cell 1999; 3(2):255-61.
19. Gradia S, Acharya S, Fishel R. The human mismatch recognition complex hMSH2-hMSH6 functions as a novel molecular switch. Cell 1997; 91(7):995-1005.
20. Fishel R. Mismatch repair, molecular switches, and signal transduction. Genes Dev 1998; 12(14):2096-101.
21. Blackwell LJ, Martik D, Bjornson KP et al. Nucleotide-promoted release of hMutSalpha from heteroduplex DNA is consistent with an ATP-dependent translocation mechanism. J Biol Chem 1998; 273(48):32055-62.
22. Allen DJ, Makhov A, Grilley M et al. MutS mediates heteroduplex loop formation by a translocation mechanism. Embo J 1997; 16(14):4467-76.
23. Obmolova G, Ban C, Hsieh P et al. Crystal structures of mismatch repair protein MutS and its complex with a substrate DNA. Nature 2000; 407(6805):703-10.
24. Lamers MH, Perrakis A, Enzlin JH et al. The crystal structure of DNA mismatch repair protein MutS binding to a G x T mismatch. Nature 2000; 407(6805):711-7.
25. Ban C, Yang W. Crystal structure and ATPase activity of MutL: Implications for DNA repair and mutagenesis. Cell 1998; 95(4):541-52.
26. Ban C, Junop M, Yang W. Transformation of MutL by ATP binding and hydrolysis: A switch in DNA mismatch repair. Cell 1999; 97:85-97.

27. Guerrette S, Acharya S, Fishel R. The interaction of the human MutL homologues in hereditary nonpolyposis colon cancer. J Biol Chem 1999; 274(10):6336-41.
28. Kondo E, Horii A, Fukushige S. The interacting domains of three MutL heterodimers in man: hMLH1 interacts with 36 homologous amino acid residues within hMLH3, hPMS1 and hPMS2. Nucleic Acids Res 2001; 29(8):1695-702.
29. Shimodaira H, Filosi N, Shibata H et al. Functional analysis of human MLH1 mutations in Saccharomyces cerevisiae. Nat Genet 1998; 19(4):384-9.
30. Ellison AR, Lofing J, Bitter GA. Functional analysis of human MLH1 and MSH2 missense variants and hybrid human-yeast MLH1 proteins in Saccharomyces cerevisiae. Hum Mol Genet 2001; 10(18):1889-900.
31. Prolla TA, Pang Q, Alani E et al. MLH1, PMS1, and MSH2 interactions during the initiation of DNA mismatch repair in yeast. Science 1994; 265(5175):1091-3.
32. Guarne A, Junop MS, Yang W. Structure and function of the N-terminal 40 kDa fragment of human PMS2: A monomeric GHL ATPase. Embo J 2001; 20(19):5521-31.
33. Guerrette S, Wilson T, Gradia S et al. Interactions of human hMSH2 with hMSH3 and hMSH2 with hMSH6: Examination of mutations found in hereditary nonpolyposis colorectal cancer. Mol Cell Biol 1998; 18(11):6616-23.
34. Heinen CD, Wilson T, Mazurek A et al. HNPCC mutations in hMSH2 result in reduced hMSH2-hMSH6 molecular switch functions. Cancer Cell 2002; 1(5):469-78.
35. Christmann M, Tomicic MT, Kaina B. Phosphorylation of mismatch repair proteins MSH2 and MSH6 affecting MutSalpha mismatch-binding activity. Nucleic Acids Res 2002; 30(9):1959-66.
36. Tran PT, Simon JA, Liskay RM. Interactions of Exo1p with components of MutLalpha in Saccharomyces cerevisiae. Proc Natl Acad Sci USA 2001; 98(17):9760-5.
37. Burdett V, Baitinger C, Viswanathan M et al. In vivo requirement for RecJ, ExoVII, ExoI, and ExoX in methyl-directed mismatch repair. Proc Natl Acad Sci USA 2001; 98(12):6765-70.
38. Jager AC, Rasmussen M, Bisgaard HC et al. HNPCC mutations in the human DNA mismatch repair gene hMLH1 influence assembly of hMutLalpha and hMLH1-hEXO1 complexes. Oncogene 2001; 20(27):3590-5.
39. Schmutte C, Sadoff MM, Shim KS et al. The interaction of DNA mismatch repair proteins with human exonuclease I. J Biol Chem 2001; 276(35):33011-8.
40. Rasmussen LJ, Rasmussen M, Lee B et al. Identification of factors interacting with hMSH2 in the fetal liver utilizing the yeast two-hybrid system. In vivo interaction through the C-terminal domains of hEXO1 and hMSH2 and comparative expression analysis. Mutat Res 2000; 460(1):41-52.
41. Genschel J, Bazemore LR, Modrich P. Human exonuclease I is required for 5' and 3' mismatch repair. J Biol Chem 2002; 277(15):13302-11.
42. Wu Y, Berends MJ, Post JG et al. Germline mutations of EXO1 gene in patients with hereditary nonpolyposis colorectal cancer (HNPCC) and atypical HNPCC forms. Gastroenterology 2001; 120(7):1580-7.
43. Bellacosa A, Cicchillitti L, Schepis F et al. MED1, a novel human methyl-CpG-binding endonuclease, interacts with DNA mismatch repair protein MLH1. Proc Natl Acad Sci USA 1999; 96(7):3969-74.
44. Hendrich B, Bird A. Identification and characterization of a family of mammalian methyl-CpG binding proteins. Mol Cell Biol 1998; 18(11):6538-47.
45. Drummond JT, Bellacosa A. Human DNA mismatch repair in vitro operates independently of methylation status at CpG sites. Nucleic Acids Res 2001; 29(11):2234-43.
46. Warbrick E. PCNA binding through a conserved motif. Bioessays 1998; 20(3):195-9.
47. Kleczkowska HE, Marra G, Lettieri T et al. hMSH3 and hMSH6 interact with PCNA and colocalize with it to replication foci. Genes Dev 2001; 15(6):724-36.
48. Ellis NA, Groden J, Ye TZ et al. The Bloom's syndrome gene product is homologous to RecQ helicases. Cell 1995; 83(4):655-66.
49. Pedrazzi G, Perrera C, Blaser H et al. Direct association of Bloom's syndrome gene product with the human mismatch repair protein MLH1. Nucleic Acids Res 2001; 29(21):4378-86.
50. Langland G, Kordich J, Creaney J et al. The Bloom's syndrome protein (BLM) interacts with MLH1 but is not required for DNA mismatch repair. J Biol Chem 2001; 276(32):30031-5.

51. Wang Y, Cortez D, Yazdi P et al. BASC, a super complex of BRCA1-associated proteins involved in the recognition and repair of aberrant DNA structures. Genes Dev 2000; 14(8):927-39.
52. Karran P, Hampson R. Genomic instability and tolerance to alkylating agents. Cancer Surveys 1996; 28:69-85.
53. Moreland NJ, Illand M, Kim YT et al. Modulation of drug resistance mediated by loss of mismatch repair by the DNA polymerase inhibitor aphidicolin. Cancer Res 1999; 9:2102-2106.
54. Yamada M, O'Regan E, Brown R et al. Selective recognition of a cisplatin-DNA adduct by human mismatch repair proteins. Nucleic Acids Res. 1997; 25:491-495.
55. Duckett DR, Drummond JT, Murchie AIH et al. Human MutSalpha recognizes damaged DNA base pairs containing O6-methylguanine, O4-methylthymine or the cisplatin-d(GpG) adduct. Proc Natl Acad Sci USA 1996; 93:6443-6446.
56. Vaisman A, Varchenko M, Umar A et al. The role of hMLH1, hMSH3, and hMSH6 defects in cisplatin and oxaliplatin resistance: Correlation with replicative bypass of platinum-DNA adducts. Cancer Res 1998; 58:3579-3585.
57. Durant S, Morris M, Illand M et al. Dependence on RAD52 and RAD1 for anticancer drug resistance mediated by inactivation of mismatch repair genes. Current Biol 1999; 9:51-54.
58. Claij N, Te Riele H. Methylation tolerance in mismatch repair proficient cells with low MSH2 protein level. Oncogene 2002; 21(18):2873-9.
59. Brown KD, Rathi A, Kamath R et al. The mismatch repair system is required for S-phase checkpoint activation. Nat Genet 2003; 33(1):80-4.
60. Gong J, Costanzo A, Yang H-Q et al. The tyrosine kinase c-Abl regulates p73 in apoptotic response to cisplatin-induced DNA damage. Nature 1999; 399:806-809.
61. Duckett DR, Bronstein SM, Taya Y et al. hMutSalpha and MutLalpha dependent phosphorylation of p53 in response to DNA methylator damage. Proc Natl Acad Sci USA 1999; 96:12384-12388.
62. Kharbanda S, Ren R, Pandey P et al. Activation of the c-Abl tyrosine kinase in the stress response to DNA- damaging agents. Nature 1995; 376(6543):785-8.
63. Nehme A, Baskaran R, Aebi S et al. Differential induction of c-Jun NH2-terminal kinase and c-Abl kinase in DNA mismatch repair-proficient and -deficient cells exposed to cisplatin. Cancer Res 1997; 57(15):3253-7.
64. Toft NJ, Winton DJ, Kelly J et al. Msh2 status modulates both apoptosis and mutation frequency in the murine small intestine. Proc Natl Acad Sci USA 1999; 96:3911-3914.
65. Zeng M, Narayanan L, Xu XS et al. Ionizing radiation-induced apoptosis via separate Pms2- and p53- dependent pathways. Cancer Res 2000; 60(17):4889-93.

CHAPTER 6

Origin, Recognition, Signaling and Repair of DNA Double-Strand Breaks in Mammalian Cells

Larry H. Thompson and Charles L. Limoli

Abstract

A chromosomal double-strand break (DSB) can arise from multiple sources including ionizing radiation and DNA replication itself. An understanding of the intricate protein pathways that recognize DSBs and recruit the DNA repair and cell cycle checkpoint machinery is developing rapidly. The ATM kinase plays an early, pivotal role in the signaling process by detecting DSBs and relaying this information to numerous downstream transducer and effector proteins. Within minutes after DSBs occur, ATM undergoes inter-molecular autophosphorylation at Ser^{1981}, which converts it to an active monomer. $ATM^{Ser1981\text{-}P}$ immediately phosphorylates histone H2AX over a megabase region of DNA surrounding a DSB. Discrete nuclear foci of phosphorylated H2AX (γH2AX) are visible by immunofluorescence and appear to be true markers of DSBs. MDC1 and 53BP1, transducer proteins that contain two C-terminal BRCT domains, are also phosphorylated by ATM and colocalize faithfully with γH2AX. Subsequent transducers and effectors include the Mre11-Rad50-NBS1 complex (both transducer and effector), and the breast cancer susceptibility proteins BRCA1 (a transducer) and BRCA2 (an effector). BRCA2 interacts directly with DNA and the Rad51 strand-transferase to help initiate homologous recombination. When the DNA replication machinery is chemically inhibited or encounters a damaged template containing single-strand breaks or blocking lesions, replication forks may arrest, collapse into one-sided DSBs, and require recombinational repair to be reestablished. This recovery process is dependent on the ATR kinase acting in concert with the Rad17-Rfc clamp-loader complex and the Rad9-Rad1-Hus1 clamp complex. Modifiers of DNA topology, such as BLM and WRN helicases associated with Bloom's and Werner's syndromes, assist in preserving chromosomal continuity during replication. These proteins are thought to resolve anomalous replication intermediates that arise at stalled forks, thereby preventing aberrant recombination for unrepaired DSBs. Overall, the precise nature of a DSB likely determines whether ATM or ATR is utilized to initiate the damage-response pathways.

Eukaryotic DNA Damage Surveillance and Repair, edited by Keith W. Caldecott.

Replication-Independent Double-Strand Breaks (DSBs)

Origins of DSBs

Double-strand breaks (DSBs) are of fundamental importance in many fields of biology. The incorrect repair of DSBs often results in chromosomal rearrangements, which are considered to be a major initiating factor in carcinogenesis. Cancer cells generally exhibit numerous structural rearrangements (i.e., deletions, exchanges, duplications, and inversions) as well as increased numbers of chromosomes. Progression of malignancy often correlates with increased chromosomal instability and plasticity, which are driven by escalating defects in DNA repair processes[1] and cell cycle checkpoint functions.[2-4] The cellular lethality of ionizing radiation (IR) occurs largely through the production of DSBs. Many cancer treatments rely on the ability of IR and chemical agents (e.g., bleomycin) to produce DSBs that can be targeted to preferentially eradicate tumor cells versus damaging normal tissues. Thus, understanding the quantitative yields of DSBs and the molecular mechanisms that eliminate them is a central issue in cancer biology and radiation biology.

The yield of breaks produced by IR is estimated to be ~35 DSBs per diploid G1 cell per Gy (measured at doses ≥ 20 Gy), compared with a value of ~1000 single-strand breaks (SSBs) per Gy.[5-8] Recent estimates of DSB yield measured by the frequency of IR-induced γH2AX foci[9,10] at doses between 0.001 and 3 Gy give a value that is very similar to the ~35 breaks per Gy determined at high doses by pulsed-field gel electrophoresis.[11] However, estimates based on premature chromosome condensation (PCC), which allows visualization of chromosomes in G_1 nuclei, are considerably lower at 5 to 6 DSB per Gy per cell.[12,13] The reason(s) for this discrepancy is unclear. One possible explanation is that IR-generated DNA fragments arising from two or more DSBs in relatively close proximity would not be microscopically distinguishable from a single DSB. Second, in the PCC method some fraction of the rapidly repaired DSBs will be missed because of the 15-20 min post-IR incubation at 37°C required to produce cell fusion and chromosome condensation. This fraction could be as high as 65%.[14] IR also produces oxidative base damage, but the amounts of damaged bases[8] per Gy of radiation are estimated to be 10-100 fold lower than the steady state levels (~1.5 x10^5 oxidative base lesions per human cell[15]) produced by normal oxygen metabolism.[16]

Recently though, it has become apparent that IR produces clustered oxidative base damages and SSBs on opposite strands of the DNA molecule, which can develop into DSBs and represent potentially lethal lesions.[17-19] Clusters of closely opposed SSBs, oxidized purines, oxidized pyrimidines, or oxidized abasic sites within a few helical turns are estimated to comprise at least 70% of the complex lesions produced in cells.[19] These clustered lesions are likely difficult to repair[20] and may get converted to DSBs through processing by base-excision repair enzymes, or by interaction with DNA replication forks as they encounter clustered lesions.

DNA Repair Systems That Act on DSBs

Cells possess complex, highly efficient mechanisms for detecting DSBs and signaling their presence to the DNA repair and replication machinery. In this review, we address what is known about these recognition and information transfer systems outlined in (Fig. 1). Our understanding of how cells respond to DSBs has developed rapidly with respect to the enzymatic machinery that performs repair. Considerably less is known about the preceding events of detecting/sensing/recognizing breaks and the signaling processes that recruit DNA repair systems to the sites of damage. The two major pathways that repair DSBs are referred to as nonhomologous end joining (NHEJ)[21,22] and homologous recombinational repair (HRR).[23-25] In this review, we address what is known about these recognition and information transfer systems, along with a brief, updated summary of HRR (Fig. 5).

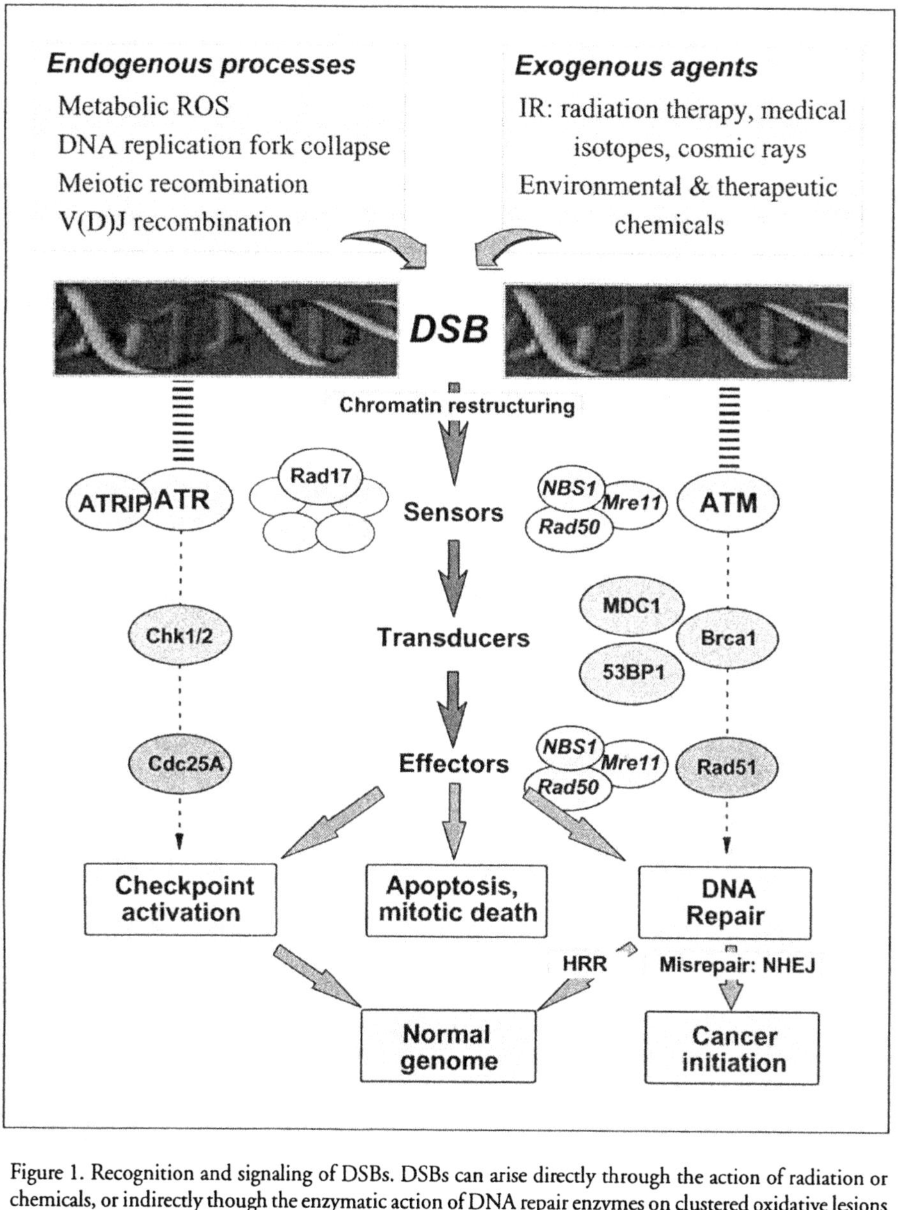

Figure 1. Recognition and signaling of DSBs. DSBs can arise directly through the action of radiation or chemicals, or indirectly though the enzymatic action of DNA repair enzymes on clustered oxidative lesions or the interaction of DNA replication forks with single-strand breaks. DSB detection is followed by signaling reactions (often phosphorylations) that implement repair and checkpoint functions, transcriptional changes, or possibly apoptosis if a cell is severely damaged. Examples of transducers are given.

DSBs occur normally during meiosis to initiate strand exchange between homologous chromosomes and in hematopoietic cells during gene processing through V(D)J recombination, which mediates antibody diversity and gene rearrangements for T cell receptors. These highly regulated, programmed DSB-mediated processes are normally extremely accurate and utilize many of the same DNA break-processing enzymes that repair spontaneous or agent-induced DSBs. High levels of spontaneous DSBs and chromosomal rearrangements are observed in mouse cells carrying null mutations in the genes of the NHEJ complexes composed of Ku70-Ku86-DNA-PKcs or LIG4-XRCC4.[26-28] The levels of chromosomal breaks can be reduced by lowering the oxygen tension from 20% to 3%.[29] Elevating the level of reactive oxidative species, rather surprisingly by overexpressing a transgene for the antioxidant enzyme superoxide dismutase 1 (SOD1), increases chromosome breakage. In SOD1-overexpressing cells, reducing oxygen to 3% also reduced chromosomal aberrations. The observation that oxidative damage results in spontaneous chromosome breaks may explain the neuronal degeneration and premature aging that typify mice having NHEJ mutations.[30-34]

However, not all NHEJ-defective cell lines (i.e., Ku70 and Ku80 mutants in hamster CHO and chicken DT40 backgrounds) display markedly increased levels of spontaneous chromosomal aberrations.[35-37] The reason for the significant differences among cell types is not clear, but it is noteworthy that both the CHO and DT40 lines are defective for Tp53. Perhaps this defect allows for increased DNA-PKcs-independent end joining in the absence of the DNA-PK or LIG4-XRCC4 complexes.

Central Role of the ATM Kinase in DSB Signaling

The large ATM (ataxia telangiectasia mutated) and ATR (AT and Rad3-related) kinases have come into focus as early, central participants in the DNA damage recognition and signaling processes (Fig. 2 and Fig. 6).[3,38-41] These functionally related proteins phosphorylate a multitude of substrates and appear to exist in vivo in high molecular weight complexes of >2 x 10^6 Da,[42,43] which may contain many other damage-response proteins.[44] Figure 2 expands the theme of (Fig. 1) by depicting numerous phosphorylation events as well as functionally important protein interactions. Exposure of cells to IR immediately activates the ATM kinase (3056 a.a.; $Tel1^{Sp}$ and $Tel1^{Sc}$ homologs in yeasts),[45,46] and ATP can also induce activation of by a mechanism involving autophosphorylation.[47] A major advance came with the discovery that IR-induced activation occurs through intermolecular autophosphorylation of Ser^{1981}, which causes dissociation of ATM dimers and enhancement of kinase activity.[48] After IR, phosphorylation of Ser^{1981} is maximal within 5 min and saturates at a dose of ~40 cGy.[48] Upon activation, ATM phosphorylates histone H2AX (modification referred to as γH2AX);[49] DNA-PK and ATR also contribute to this modification.[50-53] Although ATM binds preferentially to DNA ends in vitro,[54] the in vivo activation likely results from changes in chromatin structure instead of DNA binding.[48] ATM phosphorylates numerous key proteins that often appear in nuclear foci (described below) and that mediate checkpoints and DNA repair: namely 53BP1, MDC1/NFBD1, Chk1, Chk2, NBS1, BRCA1, and FANCD2. Altogether, ATM has more than 20 substrates, as recently reviewed in more detail.[25] Thus, throughout the cell cycle ATM acts as a master regulator and coordinator in the initial response to DSBs that are not associated with replication forks. ATM is also activated by agents such as methylating chemicals that do not directly cause DSBs, but lead to lesions that are subsequently converted to DSBs.[55] The closely related ATR kinase discussed below may serve as a partial backup system for ATM and help to reinforce at later times the phosphorylating signaling initiated by ATM (see discussion in ref. 41).

The fact that ATM is responsible for phosphorylating proteins that implement repair and checkpoint functions suggests that ATM itself might concentrate at sites of DSBs. Indeed, within 5 min after IR, the Ser^{1981}-phosphorylated form of ATM begins to form foci that colocalize with γH2AX foci, and these become distinct foci by 60 min.[48] Under conditions of detergent extraction to remove nucleoplasmic proteins, a portion of the total ATM pool becomes resistant to extraction and is detected in nuclear aggregates immediately after DSB formation.[56] These aggregates are much more diffuse than the distinct foci formed by γH2AX.

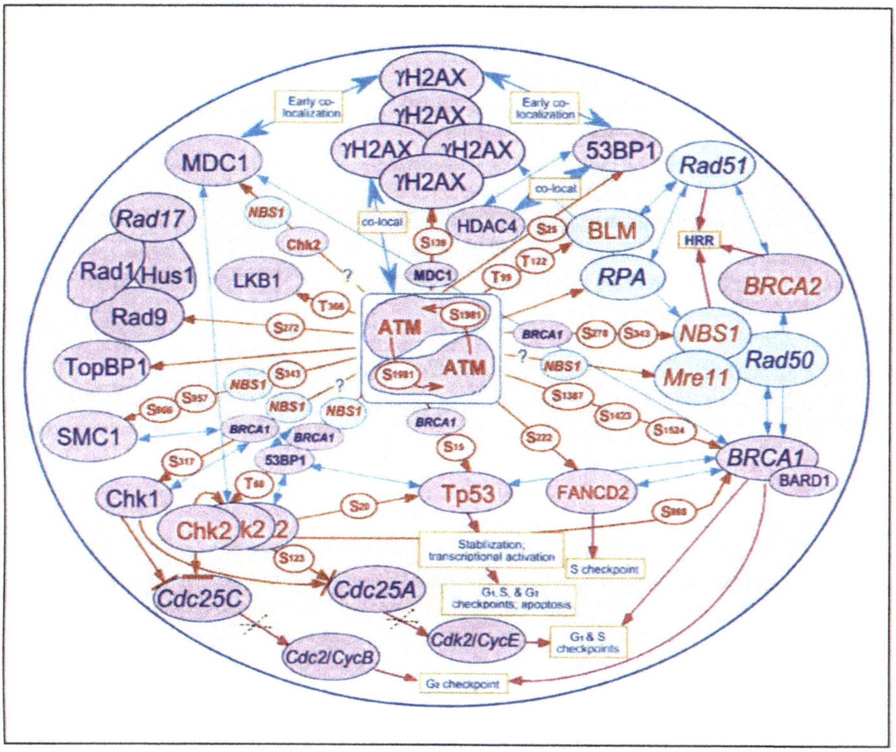

Figure 2. ATM mediated phosphorylation events and interactions that implement signaling, repair, and checkpoint functions in response to replication-independent DSBs. Within 60 sec after irradiation, phosphorylation of histone H2AX (referred to as γH2AX) occurs in discrete foci at sites of DSBs. These foci, which contain ~1000 or more γH2AX molecules in a 10^6 bp-region of DNA, likely recruit the numerous proteins that signal the presence of damage, conduct DNA repair, initiate checkpoints to halt cell cycle progression, or facilitate apoptosis in heavily damaged cells. These initial chromatin modifications must create high-affinity sites that localize such damage-response proteins. The choice of repair pathway, NHEJ or HRR, is probably dictated by the position of the cell in the cell cycle and by the structure of the DSB (e.g., one-sided versus two-sided breaks; see (Fig. 7C). A DSB occurring in a region of DNA that has replicated may be specifically channeled through HRR so that mutations can be avoided. This pathway specificity may be conferred by the proteins that are present in the vicinity of the break, as well as an altered state of the replicated chromatin, which distinguishes it from unreplicated chromatin in G1 cells and unreplicated DNA regions in S cells. Proteins with names in red are involved in human genetic disorders and proteins in italics are required for viability of dividing cells. Specific phosphorylation steps that are thought to be functionally important have been reported: H2AX,[49] BLM,[312] NBS1,[61-64] BRCA1,[222,313-317] FANCD2,[167] Tp53,[45,46,318] Chk1,[105,107] Chk2,[319-324] SMC1,[166,325] and Rad9.[326] Additional ATM-dependent phosphorylation targets are RPA,[327-329] 53BP1,[91,93] TopBP1,[127] LKB1,[330] MDC1/NFBD1,[58,79-81] and possibly Mre11.[65,331] Chk2 also phosphorylates BRCA1.[332] BRCA1 is required for phosphorylation of 53BP1, Tp53, and NBS1,[356] and NBS1 is required for the phosphorylation of MDC1, SMC1, Chk1, 53BP1, and Mre11. Other phosphorylation requirements are also indicated by proteins on the red lines. Question marks indicate that the phosphorylation could be indirect. The heavy blue arrows show proteins that colocalize within minutes after IR damage. The interactions, determined by coimmunoprecipitation, marked by blue arrows have been reported for MDC1-Chk2,[80] MDC1-BRCA1,[116] BLM-Rad51,[333] RPA-Rad51,[122] BRCA2-Rad51,[334] BRCA1-BARD1,[335] BRCA1-BRCA2,[336] BRCA1-Tp53,[337] BRCA1-Rad50,[338] 53BP1-Tp53,[87] 53BP1-Chk2,[57] 53BP1-BRCA1,[57] 53BP1-γH2AX,[91] 53BP1-HDAC4,[97] and BRCA1-FANCD2.[339]

Although the sensor proteins that first recognize DSBs are not well understood, ATM is a major candidate sensor protein. It could act alone or in combination with other proteins discussed below such as 53BP1[57] and MDC1/NFBD1[58] that localize within minutes to sites of DSBs. It is noteworthy that certain ATM mutations display a dominant negative phenotype in the heterozygous state, both in humans and mice.[59,60] This situation could arise if mutant ATM binds and sequesters partner proteins into dysfunctional complexes that compete with normal complexes for DNA substrates. Although ATM is required to phosphorylate both NBS1 and Mre11,[61-65] genetic evidence suggests that the Mre11-Rad50-NBS1 (MRN) complex acts upstream of ATM, at least for some signaling events.[66] Mre11-defective human cells show reductions in detergent-resistant retention of ATM protein, ATM kinase activity, and phosphorylation of downstream targets.

Origins of Nuclear Foci That Form in Response to DSBs

When cells are exposed to DNA damaging agents, the redistribution and subnuclear localization of specific proteins can be monitored to infer which proteins are important in damage recognition, signaling, checkpoint implementation, and repair. In mammalian cells, Rad51 protein, involved in homologous recombination, was one of the first proteins detected in discrete nuclear foci using immunofluorescence on mitotic and meiotic cells[67,68] (Fig. 3A). The cytological visibility of these foci can be readily explained by the fact that Rad51 forms nucleoprotein filaments that can contain hundreds of Rad51 molecules. It is estimated that ~100 fluorophore molecules localized within a very small volume are necessary for a visible focus.[69] Many other proteins discussed below also form foci, but not all of these are expected to assemble en masse as multimeric functional complexes like Rad51. For example, the MRN complex is a key component in the processing of DSB termini,[70-72] and the MRN complex forms foci. Although the precise biochemical roles of this complex are not understood, only one or a few of these complexes, as a catalytic component, may be needed to produce single-stranded tails at the termini of DSBs prior to their repair by homologous recombination. Yet, focus formation may arise from the creation of multiple high-affinity binding sites for MRN in the vicinity of the DSB, thus causing numerous MRN complexes to concentrate at the modified site. This idea is illustrated by the observation that foci of Rad52 and Rad54 are highly dynamic structures.[73] These foci exhibit rapid exchange of these proteins, which are recruited independently with differing mobility.

γH2AX Formation As a Marker of Radiation-Induced DSBs: Impact on Checkpoints and Repair

Recent developments suggest the possibility that units of higher order chromatin structure may facilitate the detection of DSBs within DNA. In response to the introduction of DSBs, the minor histone H2AX is rapidly phosphorylated in a dose-dependent manner on Ser^{139} (located four amino acids from the carboxyl terminus), yielding the form designated γH2AX.[49] Each DSB produces ~2000 γH2AX molecules and results in the modification of H2AX over a region corresponding to ~2 Mbp,[49] which is the equivalent of 0.03% of the chromatin. An antibody specific for the modified C-terminus of H2AX reveals that γH2AX appears as discrete nuclear foci within 1 min after exposure of cells to ionizing radiation.[50] Cells in all phases of the cycle, including mitosis, show foci.[50,53] Importantly, the number of foci agrees with the estimated number of induced DNA DSBs.[10,11,50] Significant phosphorylation of H2AX was also observed after treatment with the DSB-inducing agents neocarzinostatin, bleomycin, and etoposide, whereas UV irradiation and the DNA methylating agent methyl methanesulfonate did not produce γH2AX.[51] This pattern further supports the idea that γH2AX phosphorylation occurs specifically in response to DSBs. Recent studies show that the formation of γH2AX is severely reduced in DNA-damaged ataxia telangiectasia cells and that the

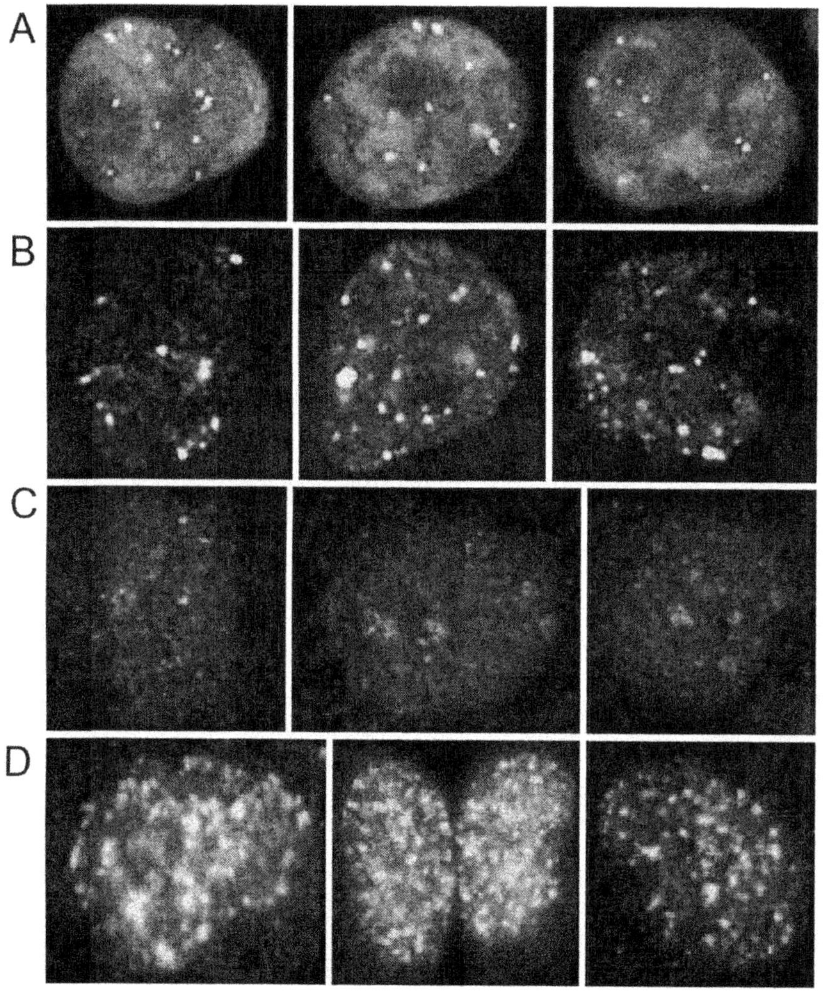

Figure 3. Rad51 and γH2AX nuclear foci produced by ionizing radiation. (A) CHO AA8 cells irradiated with 10 Gy and stained with Rad51 antibody and DAPI after 4 h. (B) Unirradiated CHO AA8 cells stained with γH2AX antibody. Putative S phase cells are shown; many cells had no foci. A recent study suggests that only the most intense foci represent true DSBs,[134] which emphasizes the importance of calibrating the scoring of foci. (C) Unirradiated GM637 fibroblasts stained with γH2AX antibody. (D) GM637 cells treated with 6 Gy and stained with γH2AX antibody after 30 min.

residual level could be attributed to DNA-PK activity [51,74] NBS cells (Nijmegen Breakage Syndrome; NBS1 is part of the MRN complex) have normal γH2AX focus formation,[75] but depletion of the MDC1 signal transducer discussed below reduces focus formation.[58] IR-induced H2AX phosphorylation is a highly conserved process that is present in vertebrates, Drosophila, and yeast.[50] γH2AX is formed in response to DSBs arising by diverse means: directly from environmental insult by radiation or chemicals, collapse of DNA replication forks, and programmed processes that enzymatically introduce DSBs (e.g., meiosis). The trigger for γH2AX formation may involve topological changes in the DNA, such as the degree of super coiling.

The biological importance of γH2AX formation in maintaining chromosome stability is clearly revealed by the phenotypes of mice carrying knockout mutations in one or both copies the *H2AX* gene.[76,77] *H2AX*$^{\Delta/\Delta}$ mice are growth retarded, radiation sensitive, immune deficient, and defective in spermatogenesis. The IR sensitivity of *H2AX*$^{\Delta/\Delta}$ ES (embryonic stem) cells is increased ~3-fold,[77] whereas the sensitivity of immortalized MEF (mouse embryonic fibroblast) cultures is increased only ~1.6-fold, but nevertheless these latter cells were shown to have reduced DSB repair.[76] Spontaneous chromosomal aberrations are also markedly elevated, e.g., from 5% in heterozygous *H2AXFlox/*$^{\Delta}$ controls to 22% in *H2AX*$^{\Delta/\Delta}$ ES cells.[77] Although checkpoint functions in all phases of the cell cycle were considered to be normal in *H2AX*$^{\Delta/\Delta}$ cells following 10 Gy irradiation,[76,77] at lower doses a clear G2 checkpoint defect was seen in both mouse B cells and MEFs.[78] Thus, H2AX phosphorylation signals for both checkpoint activation and repair. In summary, the pleiotropic phenotype of *H2AX*$^{\Delta/\Delta}$ mice is caused by defects in signaling that include impaired recruitment of MDC1, 53Bp1, NBS1, and Brca1 (but not Rad51) into IR-induced foci.[58,76,77]

Recruitment and Colocalization of Signaling and Repair Proteins to Sites of γH2AX Foci

In this section we outline the characteristics of many proteins that are implicated in signaling, checkpoints, and repair through their redistribution within the nucleus in response to DNA breakage. Some of these proteins, such as MDC1/NFBD1 and 53BP1, appear to arise as quickly as, and coincident with, γH2AX foci while other foci (e.g., Rad51) arise much later. By further example, colocalizing γH2AX-BRCA1 foci were reported to appear sooner than γH2AX-Rad50 foci,[75] implying that BRCA1 may act upstream of the Rad50 complex (MRN, discussed below). Figure 4 summarizes available information on the order of appearance of foci, including colocalizations. Certainly there are numerous pitfalls in deciphering the significance of focus formation. Foci studies performed at high doses (e.g., > 5 Gy) and many hours after exposure will be much more difficult to interpret than studies terminated minutes after irradiation at low doses (e.g., < 1 Gy). The remainder of this section summarizes the characteristics and significance of many proteins that have been shown to form nuclear foci.

MDC1/NFBD1

Very recently a new nuclear human protein, MDC1 (mediator of DNA checkpoint; also called NFBD1 for nuclear factor containing two BRCT domains at the C-terminus; 2089 a.a.), which constitutively binds to chromatin, was identified as a very early participant in the recognition and signaling process.[58,79-82] Because of its C-terminal BRCT domains, MDC1 is a candidate functional homolog of Rad9^{Sc}, one of the first checkpoint proteins to be identified in budding yeast.[83,84] After IR damage, MDC1 becomes hyperphosphorylated in an ATM-, NBS1-, and Chk2-dependent manner,[58,80,81] but MDC1 focus formation is still seen in the absence of ATM.[81] Within 1 min after irradiation, MDC1 forms visible foci that peak in frequency at 30 min.[58,82] MDC1 foci colocalize precisely with γH2AX foci, and γH2AX is needed for MDC1 focus formation.[58,82] H2AX and MDC1 are mutually interdependent for phosphorylation and focus formation, and MDC1 forms complexes with γH2AX.[58] However, there are conflicting data concerning whether MDC1 is required for 53BP1 focus formation.[5882] Suppression of MDC1 by siRNA results in decreased phosphorylation of SMC1^{S996} and Chk1^{S345}.[58] This defect in Chk1 phosphorylation results in defective intra-S phase and G2 checkpoints after IR exposure. Suppression of MDC1 expression by siRNA also causes reduced apoptosis in response to IR damage (because of the loss of the MDC1-Chk2^{P68} interaction[80]), but also decreases colony-forming ability.[82]

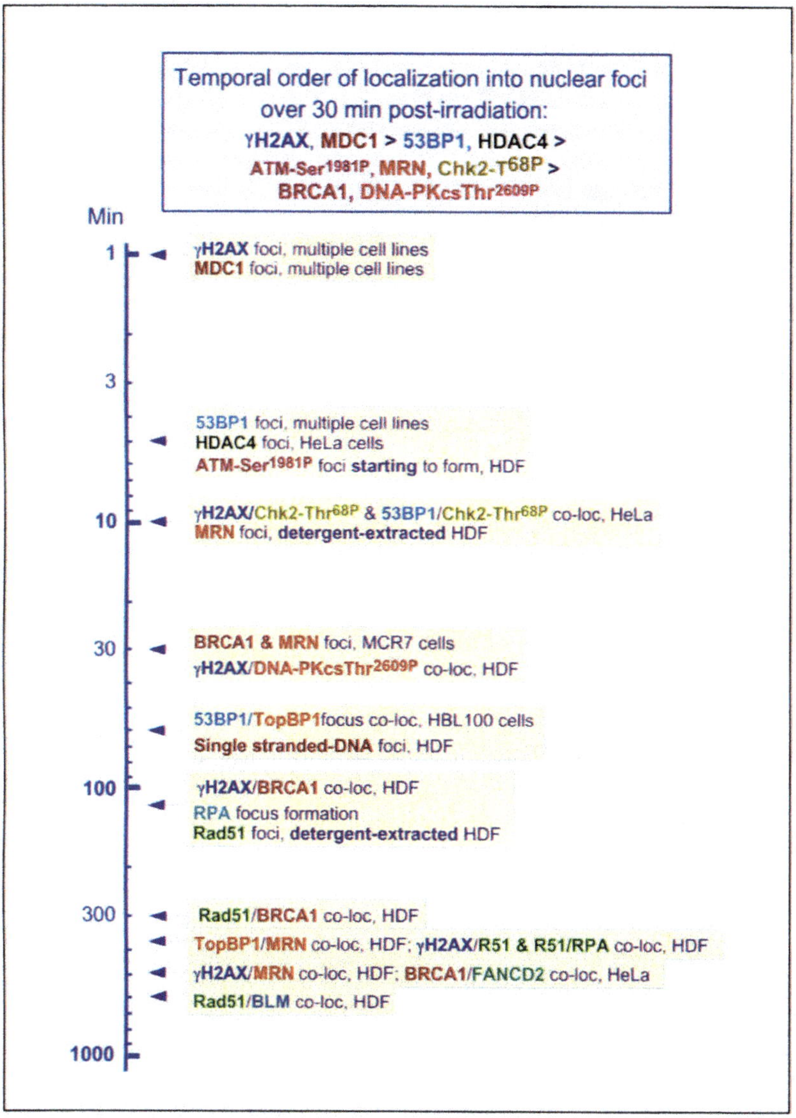

Figure 4. Timeline of recruitment of signaling, repair, and checkpoint proteins to sites of DSBs. Immediately after irradiation the formation of phosphorylated histone H2AX (γH2AX) is detectable and appears to be a/the critical event that modifies chromatin and provides an expansive high affinity site to recruit the repair and checkpoint proteins. The approximate order of appearance of key proteins that have been identified in nuclear foci is indicated. In some instances the position of the arrow may be the earliest time point that was reported, rather than the time of first appearance of the foci. The time of greatest abundance of the foci is often later than when they first appear. The kinetics of appearance of foci containing the MRN (Mre11-Rad50-NBS1) complex or Rad51 is much more rapid in detergent extracted cells.[114] Many studies were done using IR doses in excess of 2 Gy and represent observations on nonsurviving cells. Particularly at later times, foci may represent abortive repair events in dying cells. Most studies have used asynchronous cell populations. Additional work with synchronized cells will provide more precise descriptions of the order of events.

53BP1

Human 53BP1 (1972 a.a. and containing two C-terminal BRCT[85] repeats like MDC1/NFBD1) was identified as a Tp53-interacting protein in a yeast two-hybrid screen[86,87] and was found to have homology with Rad9^{Sc}.[88] 53BP1-null (or 53BP1-truncated) mice have a phenotype resembling that of H2AX-deficient mice and that of AT, i.e., growth retardation, immune deficiency, radiation sensitivity, impaired Chk2 phosphorylation/activation (Fig. 2), and cancer proneness.[89,90] Although embryo-derived cultures appear to have intact G1, S, and G2 checkpoints,[89] other studies with different cell types report a requirement for 53BP1 in the S and G2 checkpoints.[57,78]

Cytologically, 53BP1 shows diffuse nuclear immunostaining in undamaged G1 cells but a punctate pattern in S phase,[90] suggesting localization to sites of stalled or broken replication forks. 53BP1 localizes into nuclear foci in most cells as early as five min after IR exposure; doses as low as 0.5 Gy result in formation of these foci.[88] After 1 Gy, the maximal numbers of foci per cell (≈ 20-35) and of foci-positive cells (~92%) are seen between 15 and 120 min.[78,88,91] This number of foci is consistent with the expected yield of DSBs, as discussed above. Significant colocalization of 53BP1 with γH2AX is seen between 10 and 240 min,[78,88,91] and the two proteins show damage-dependent coimmunoprecipitation.[91] As with γH2AX, 53BP1 focus formation is specific for agents that produce DSBs and occurs more rapidly than that of other proteins discussed below (BRCA1, MRN, and Rad51). Phosphorylation and focus formation of 53BP1 are controlled independently.[92] One notable difference between 53BP1 and γH2AX is that only γH2AX forms foci in mitosis,[50,93] where 53BP1 associates with kinetocores.[94]

Recent work shows that 53BP1 has a central role in IR-induced DSB signaling for the S- and G2-phase checkpoints [see editorial in ref. 95]. Upon inhibition of 53BP1 by siRNA, phosphorylation of Tp53 and BRCA1 by ATM is blocked, and the downstream formation of IR-induced BRCA1 foci is largely abolished.[57] Studies with mouse knockout cells show that 53BP1 is required for a normal G2-M checkpoint at low IR doses.[78] In AT cells, 53BP1 focus formation after 1 Gy was diminished at early times (10-20 min post irradiation) in one study,[91] and the hyperphosphorylation that occurs after irradiation was absent or reduced.[91,93] However, in other studies that used 3 or 8 Gy, 53BP1 focus formation appeared to be normal in AT cells,[88,96] perhaps because of the higher doses used. ATM phosphorylates 53BP1 in vitro,[78,91,92] further suggesting that ATM is responsible for directly phosphorylating 53BP1 within γH2AX foci. ATM and 53BP1 also show IR-dependent coimmunoprecipitation.[96] 53BP1 may either help recruit and activate ATM at sites of DSBs, or recruit ATM substrates to these sites.[96] It is noteworthy that 53BP1, as well as Chk2, appears to be present in γH2AX foci before any of the proteins that participate in DSB repair. These results suggest that the initiation of checkpoints precedes the onset of DSB repair, which occurs over a period of several hours, depending on the dose.

HDAC4

Histone deacetylase 4 (HDAC4) is another early participant in IR-induced focus formation, and HDAC4 foci are seen in cells defective in ATM, DNA-PKcs, or NBS1.[97] In contrast, HDAC2 or HDAC6 do not form foci. Interestingly, the stability of 53BP1, as well as its focus formation, shows a dependence on the presence of HDAC4, as shown by siRNA inhibition experiments. Depletion of HDAC4 abrogates the G2 delay and confers sensitization to killing by IR while also reducing cell viability.[97] In this study the authors suggested that the degree of persistence of HDAC4 foci might be a measure of cellular radiosensitivity.

Chk2

Chk2/Cds1 is a key checkpoint kinase (for reviews see refs. 3,98,99), whose role appears to be primarily promoting apoptosis, not cell survival, after IR exposure.[100,101] Chk2-deficient mice are radioresistant and defective in Tp53-mediated transcriptional changes.[100,101] Thymocytes, splenocytes, skin, and neurons in the developing brain show protection from IR-induced apoptosis. The G1 checkpoint, but not the G2 or S-phase checkpoints, was substantially impaired in *Chk2-/-* embryonic fibroblasts and ES cells.[100,101] IR-induced stabilization of Tp53 in *Chk2-/-* cells is ~60% of that in wild-type cells.[101] Caffeine further reduces Tp53 accumulation, suggesting the presence of another pathway for Tp53 stabilization that is ATM/ATR-dependent, but Chk2-independent. In spite of Tp53's partial stabilization and phosphorylation at Ser^{23} (Ser^{20} in human cells) in the absence of Chk2, Tp53-dependent transcriptional induction of target genes, such as *CDKN1A/p21*, was not observed in *Chk2-/-* cells.

IR treatment results in ATM-dependent activation and phosphorylation of Chk2 at Thr^{68}.[102] The complete activation of Chk2 requires NBS1.[103,104] Immuno-staining using phospho-specific Chk2 antibodies suggests that Chk2-Thr^{68P} localizes into discrete foci within 10 min after irradiation and colocalizes with both γH2AX and 53BP1 (Fig. 4).[57,102] These Chk2-Thr^{68P} foci did not appear in AT cells or in 53BP1-depleted cells, and nonphosphorylated Chk2 molecules remain distributed throughout the nucleus.[57] However, a subsequent study using 53BP1 mouse knockout cells and a different Chk2-Thr^{68P} antibody concluded that 53BP1 is not required for Chk2 activation.[78] These different conclusions might be accounted for by inadequate specificity of the Chk2-Thr^{68P} antibody[102] used in the former studies (see commentary in ref. 95). In summary, phosphorylation and focus formation of Chk2 may be a major determinant in programming cells for elimination by apoptosis through Chk2's site-specific phosphorylation of Tp53.

Chk1

Few Chk1 foci studies have been reported, but Chk1 and BRCA1 foci colocalize in the absence of IR exposure.[105] In comparison to Chk2 described above, after IR damage in chicken DT40 cells Chk1 promotes reproductive survival, apoptosis (in these Tp53-deficient cells), implementation of the G2 checkpoint, and phosphorylation of Cdc2.[106] Notably, *Chk1* null cells completely lose the G2 checkpoint in DT40 cells (see Fig. 2). Inhibition of human Chk1 with siRNA also results in a G2 checkpoint defect, and phosphorylation of Ser^{317} and Ser^{345} appears nonessential for IR-mediated activation of Chk1 and the G2 checkpoint.[107] IR activation of Chk1 depends on BRCA1.[105]

MRN Complex

The analysis of nuclear focus formation by various damage-response proteins suggests that γH2AX formation plays a critical role in recruiting and assembling repair proteins, including the Mre11-Rad50-NBS1 (MRN) complex. This complex is essential for HRR in human cells since hypomorphic mutations confer radiosensitivity.[108-110] MRN localizes to sites of damage within 30 min after irradiation[111] (Fig. 4), and recent genetic evidence suggests that it may be a primary DSB recognition factor.[66] By using a 390-nm laser combined with BrdUrd incorporation and Hoechst dye 33258[112] to produce striped regions of DSBs, Bonner and coworkers have shown in human breast tumor MCF7 cells that γH2AX stripes appear rapidly in all cells, and MRN colocalizes to these stripes within 30 min.[75] Colocalization of NBS1 and γH2AX foci appears to involve a direct interaction between NBS1 and γH2AX (and not H2AX), which is mediated by the FHA/BRCT domain of NBS1.[52]

Initial observations of MRN foci indicated that the kinetics of appearance was too delayed to correspond with productive repair of DSBs.[113] However, the patterns of focus formation of proteins such as Rad50 and Rad51 are strongly influenced by the method of preparing the cells. In earlier experiments using methanol fixation and acetone to permeabilize human diploid fibroblasts, the percentage of nuclei that were positive for Rad50 focus formation after 12 Gy γ-irradiation reached a maximum of ~65% after 8 hr.[113] In comparison, nuclei with Rad51 foci reached a broader maximum between 4-8 hr at ~35%. Subsequent studies in detergent extracted cells, designed to reveal proteins tightly associated with chromatin, found that MRN foci were detectable within 10 min after irradiation and reached a maximum at 2 hr when 70-95% of cells were positive.[114] Since MRN foci are present in AT cells, MRN focus formation does not depend on the phosphorylation of NBS1 by ATM.[61-64] These results suggest that the MRN complex may participate in a very early, ATM-independent step of DSB recognition.[66] Larger aggregates of MRN foci that become apparent by 8 hr in normal cells were not seen in AT cells.[114] These larger, more robust foci may represent sites of slow or abortive repair of complex DSBs. In unirradiated detergent-extracted cells, Mre11 shows a high degree of colocalization with immunostaining of PML bodies, a nuclear depot of many proteins that may help regulate cellular defense against insults such as viruses.[115]

BRCA1 and Rad51

Radiation-induced γH2AX focus formation occurs within several minutes in both MCF7 and IMR90 human cells after exposure to only 0.6 Gy.[50] However, higher radiation doses are usually needed to visualize the foci formed by several key proteins that are recruited to the γH2AX foci. After 12 Gy of IR, the kinetics of BRCA1 focus formation is significantly more rapid than that of MRN focus formation.[75] By 2 hr, 10-15% of IMR90 cells show BRCA1 foci, and these overlap extensively with γH2AX foci.[50] While γH2AX-BRCA1 colocalization is maximal by 2 hr, γH2AX-MRN colocalization increases up to 8 hr. BRCA1 focus formation and hyperphosphorylation after IR is dependent on MDC1.[116] In the absence of detergent extraction, Rad51 foci appear at about the same time as γH2AX-MRN foci, but in different cells.[113] Since MRN acts upstream of Rad51 (see Fig. 5), this observation is paradoxical unless all DSB repair events in a given cell were to occur in a synchronous manner. Rad51 foci become prominent at 6 hr in 20-25% of the cells, when the majority of these foci colocalize with BRCA1 foci. Rad51 foci form relatively slowly, are much less numerous than the estimated numbers of DSBs, and require high doses for their detection. Therefore, their biological significance is still unclear.

In SV40-transformed fibroblasts, the kinetics of focus formation was faster (e.g., γH2AX-BRCA1 colocalization in 45 min) and more cells had Rad51 foci, but the order of foci appearance was the same as for IMR90 cells.[75] The relatively early appearance of γH2AX-BRCA1 foci suggests that BRCA1 might interact directly with DNA breaks.[117]

Pretreatment of MCF7 breast carcinoma cells for 30 min with phosphatidylinositol-3-kinase inhibitor wortmannin, which inhibits ATM and DNA-PKcs, completely blocks γH2AX focus formation after irradiation and prevents the formation of repair-protein foci.[75] Importantly, when added 5 min after irradiation wortmannin has no effect on γH2AX, BRCA1, and Rad51 focus formation. These results emphasize that γH2AX is an early, critical event that initiates DNA repair processes. Consistent with this idea is the finding in mouse *H2AX*$^{\Delta/\Delta}$ knockout B cells or ES cells that BRCA1 and MRN foci cannot form in response to irradiation.[76,77] However, Rad51 foci do form in both these mutant cell types although their intensity seems to be diminished in the mutant ES cells exposed to 20 Gy.[76,77]

DNA-PK

The kinase activity of DNA-PK is required for the efficient repair of DSBs by NHEJ, but the molecular mechanism underlying this activity is not understood.[118,119] Very recently the catalytic subunit of DNA-PK (DNA-PKcs) was found to undergo autophosphorylation in vitro at the highly conserved Thr^{2609} and phosphorylation at this position occurred in vivo in response to IR damage.[120] DNA-PKcs ^{Thr2609} colocalizes within 30 min after IR with both γH2AX and 53BP1. After 10 Gy the phosphorylation of Thr^{2609} is maximal by 30 min and persists for up to 4 hr. The biological significance of this phosphorylation is indicated by the

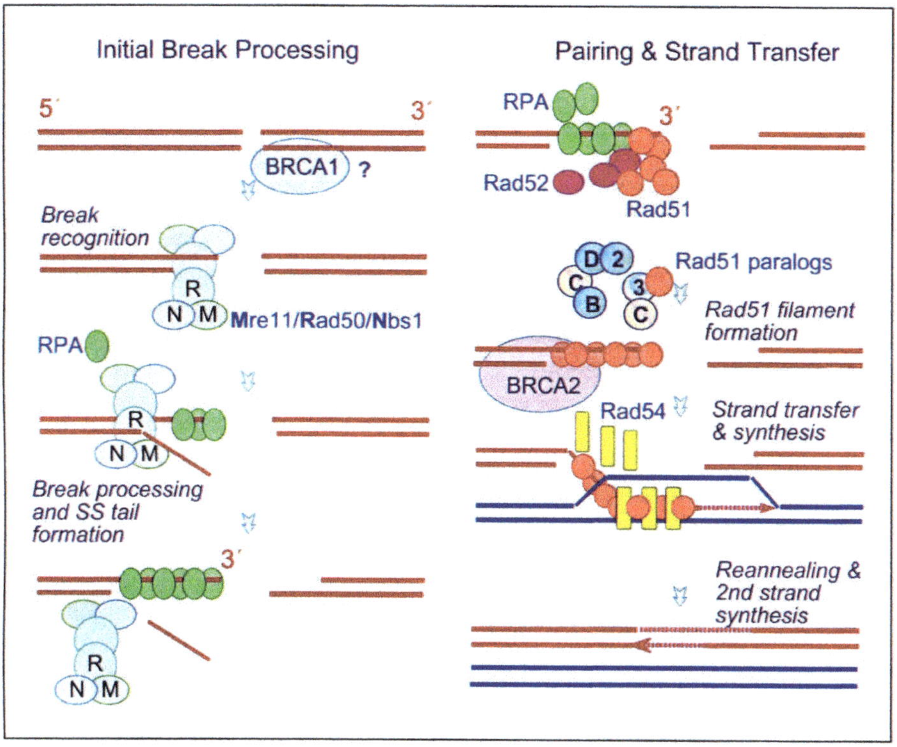

Figure 5. Outline of events in HRR. BRCA1 has an important, undefined role in regulating HRR. The MRN complex likely participates in both the recognition and processing of DNA ends to produce single-stranded tails, which will bind avidly to RPA. Formation of the Rad51 nucleoprotein filament appears to involve the action of BRCA2, the five Rad51 paralogs, and Rad52. Rad52 has functional redundancy with XRCC3,[340] and possibly other Rad51 paralogs.

finding that a Thr2609Ala substitution mutation is associated with defective DSB rejoining and causes increased radiation sensitivity in CHO cells. DNA-PK autophosphorylation at $Thr^{2609}$is reduced in the absence of ATM. Attempts to identify Ku70/86 foci in irradiated cells have generally been unsuccessful in seeing localization of this DNA end-binding complex.[114,121] Immunostaining of Ku70/86, or total DNA-PKcs, in the absence of detergent extraction, does not reveal discrete foci. This specificity suggests that only $\text{DNA-PKcs}^{\text{Thr2609}}$ may be needed to recruit Ku70/86 to DSBs, and that a high local concentration of Ku70/86 at a break may be unnecessary because of its abundance in the nucleus.

RPA

The trimeric RPA complex binds single-stranded DNA and is an essential component of both HRR and DNA replication. RPA might provide an appropriate marker for foci in which recombination has been initiated through the processing of DSB ends into structures containing single-stranded DNA coated with RPA. In human cells, interaction between RPA and Rad51 is mediated by the 70-kDa subunit of RPA.[122] RPA foci induced by irradiation are detectable at doses as low as 0.5 Gy and are present at 2 hr (and possibly earlier).[123] In focus-positive cells, the number of RPA foci per nucleus reached a maximum at ~3 hr. It would be of interest to examine the colocalization of RPA with Rad51 at low doses that allow relatively high cell survival (e.g., 0.5 Gy). This information might allow an estimate of the time required for Rad51 filament formation to occur.

TopBP1

Human TopBP1 (1522 a.a.) has sequence similarity with other checkpoint proteins (Rad4/Cut5^{Sc}, Dpb11^{Sp}, & Mus101^{Dm}) and contains eight BRCT domains.[124] TopBP1 binds to DNA ends through its BRCT domains,[125] is required for DNA replication, and interacts with Polε.[126] After radiation damage, ATM phosphorylates TopBP1, but TopBP1 focus formation occurs even in AT cells.[127] TopBP1 colocalizes ~50% with 53BP1 within one hr after irradiation and substantially colocalizes with NBS1 and BRCA1 at six hr after irradiation (Fig. 4). Inhibition of DNA synthesis by hydroxyurea results in relocalization of TopBP1 together with BRCA1 to replication forks, suggesting a role for TopBP1 in rescue of stalled forks.[126]

Kinetics of DSB Repair and Contributions of NHEJ versus HRR

It is informative to compare the information on redistribution of repair proteins in (Fig. 4) with the published studies on the kinetics of DSB rejoining using pulsed-field gel electrophoresis. Rejoining experiments are generally conducted at even higher radiation doses than those used in foci studies. Therefore, the kinetics observed may underestimate the rates of DSB repair occurring at physiological levels of damage. Nevertheless, in a variety of vertebrate cell lines, DSB repair measured by electrophoresis or neutral filter-elution exhibits a rapid component with a half-life of ~15 min and a slow component of ≥ 3-5 hr.[14,128,129] SSBs are repaired, more rapidly than DSBs, and with a rapid component that has a half-life of ~ 4 min.[130,131] In a study using premature chromosome condensation to monitor the rate of repair of visible chromosomal breaks after 6 Gy of x-rays, normal fibroblasts eliminated breaks with a half-life of 1.7 hr,[12] which is intermediate between the rapid and slow components determined by electrophoresis.

The analysis of mutant cell lines provides insight into the relative contributions of NHEJ and HRR pathways to the kinetics of DSB repair. In mouse *scid* cells, which have mutant DNA-PKcs, primarily the slow component of the biphasic curve was prolonged after a high IR dose. The final residual level of breaks was the same as in control cells.[132] Similarly, an extensive study of NHEJ and HRR mutants of chicken DT40 cells found that the *ku70* mutant, but not HRR mutants (*rad51*, *rad52*, *rad54*, and *rad51b*), had an increased half-time for DSB rejoining.[14] However, since assays of DNA size typically do not distinguish between correctly and incorrectly rejoined ends, there could well be qualitative differences between the mutants in the two pathways. Correct rejoining events, measured by restriction-fragment analysis in normal human fibroblasts, occur primarily within the first two hours, and misjoining events occur more slowly.[133] Recent work indicating the saturation of HRR at high IR doses (e.g., 20 Gy)[134] suggests that DSB repair studies done at such doses have limited biological relevance. This saturation explains why HRR mutants have not shown defects in physical assays of DSB repair.

HRR As an Error-Free Mechanism of DSB Repair in S and G2 Phases

HRR acts on DSBs that arise at broken replication forks or on DSBs occurring in segments of DNA that have already replicated. In the G1 phase of the cell cycle, homologous chromosomes do not participate in HRR at an appreciable frequency.[135] Since the proteins that mediate HRR in mammalian cells were recently reviewed,[24,25] they will not be discussed in detail. Figure 5 outlines the steps in HRR. BRCA1 and BRCA2 are essential for efficient HRR,[136,137] but the precise biochemical role for BRCA1 remains unclear. The precise functions of the MRN complex are also poorly understood. It may have architectural functions besides an enzymatic role in end processing as shown in (Fig. 5)[72,138-140] The finding that null mutations in any of the MRN proteins, as well as in Rad51 and BRCA1/2, are incompatible with cell viability (see review in ref. 25) points to their having essential functions that likely coordinate HRR with DNA replication. (This feature of essentialness also imposes limitations in determining the quantitative contribution of HRR to DSB repair.) Hypomorphic mutations in BRCA2 are remarkably similar in phenotype to mutants of each of the five Rad51 paralogs (XRCC2, XRCC3, Rad51B, Rad51C, and Rad51D) (see review in ref. 25). Recent structural studies implicate BRCA2 directly in binding single-stranded DNA and in assembling the Rad51 nucleoprotein filament.[141-143]

Replication Associated DSBs (One-Sided Breaks)

Overview

As cells proliferate, DNA-damaging biochemical reactions produce lesions that interact with the DNA replication machinery. Electron transfer during oxidative phosphorylation produces reactive oxidative species, which generate DNA single-strand breaks and oxidation products; enzymatic activation of procarcinogens generates species that form bulky adducts. Chromosomal discontinuities will arise during S-phase more frequently when replication encounters lesions (SSBs, adducts, oxidized bases, or abasic sites). Thus, a wide range of DNA damages likely give rise to DSBs during DNA replication. To deal with replication-interfering damages, cells possess an impressive array of safeguards that begin with the informational redundancy in the DNA duplex. Multiple, specialized polymerases with lesion-bypass activities help maintain the integrity of the DNA molecule,[144-146] but these error-prone polymerases have a finite coping capacity. Elegant interrelated checkpoint and repair systems are highly integrated with the replication and transcription machinery to prevent broken or rearranged chromosomes from being passed to daughter cells.

A variety of recent studies provide compelling evidence that DSBs normally arise during DNA replication. *First*, null mutations in ATR[147] or the homologous recombination machinery (*mre11*, *nbs1*, *rad50*, and *rad51*mutants) result in cell lethality that is associated with extensive chromosome breakage at metaphase.[148-153] In sperm nuclei replicating in Xenopus egg extracts, DSBs are detected as ends that label with terminal transferase and by the formation of γH2AX.[154] These breaks are only detectable when DNA replicates in the absence of the MRN complex, indicating a vital role for MRN in repairing replication-associated breaks. Since low levels of γH2AX foci are normally present in S phase mammalian cells (less than one visible focus per cell in one study),[134] and γH2AX foci directly correlate with DSBs,[11] S-phase DSBs are likely rapidly repaired. *Second*, an extrapolation of findings in *S. cerevisiae*,[155] based on relative genome size, indicates that ~100 homologous recombination events, which would be initiated by DSBs, might occur in a diploid mammalian cell during each S phase. *Third*, this numerical estimate is similar to that (i.e., >90) derived from the frequency of sister-chromatid exchange and the very low frequency of crossing over during HRR of endonuclease-generated DSBs.[25]

DSBs arising at replication forks are thought to trigger checkpoint signaling by ATR[156,157] and are dealt with by replication-fork restart and recombinational repair mechanisms (see reviews in refs. 158,159). ATR's burden of maintaining chromosome continuity during replication becomes heavier in cells lacking Tp53 (e.g., many kinds of tumor cells) because of a defective G1 checkpoint, which normally allows for the removal of damage before DNA replication.[160] It should be emphasized that DSBs arising at replication forks differ topologically from those produced by IR in that they generally involve the creation of only one double-stranded end, i.e., a one-ended chromosome break.[161,162] Such asymmetric DSBs may be preferentially recognized by the HRR machinery to accomplish error-free repair, with a lesser role played by the NHEJ machinery during S phase.

This section deals primarily with DSBs that are associated with replication and elicit the "replication checkpoint", triggered by abnormal DNA structures arising as a consequence of a blocked or collapsed (broken) fork. However, another checkpoint pathway in S phase cells has been defined historically and is referred to as the "intra-S" or "S-phase" checkpoint, which occurs when cells are exposed to IR and the DSBs are not fork-associated. When the S-phase checkpoint is activated, the initiation of replication is preferentially inhibited compared with elongation of active replicons. The signal for this inhibition is presumably DSBs although the finding that deficiencies in Msh2 and Mlh1 compromise this checkpoint[163] prompt the question as to whether other DNA lesions may initiate the process. The S phase checkpoint was originally identified by the finding that AT cells displayed "radioresistant DNA synthesis" caused by the lack of inhibition of replicon initiation.[164] Besides ATM,[165] the S-phase checkpoint requires MDC1,[58,80,81] NBS1,[61,70,108] Mre11,[109] Chk1,[107] SMC1,[166] FANCD2,[167] and Msh2/Mlh1.[163] Two complementary pathways for S-phase checkpoint activation, both of which require ATM, have been described. One subpathway operates through the Chk2 kinase and the other through phosphorylated NBS1 and SMC1 (see Fig. 2, lower left).[166,168] The function of phosphorylated SMC1 is not yet known. Thus, the replication and S-phase checkpoints both act to slow the progression of cells through S phase, but act through different pathways (compare Fig. 2 and Fig. 6). The replication checkpoint depends primarily on ATR rather than ATM, as diagrammed in (Fig. 6) Despite these differences, the two pathways exhibit overlap in the activation of their downstream effectors. After IR damage, the G2/M checkpoint requires the cooperation of both ATM and ATR, as revealed from an elegant analysis of single and double mutants.[169]

Replication-Associated DSBs Arising from a Damaged Template

In the simplest case, a DSB may occur in one daughter chromatid when a replication fork encounters a SSB, which can arise as an intermediate in base excision repair. Under conditions where excess SSBs are present, as during the repair of methylation damage (e.g., MMS exposure), the production of replication-associated DSBs will be exacerbated. When helicase and polymerase activities become uncoupled at the replication fork[170-172] and generate extended regions of single-stranded DNA (e.g., 1 kb), the likelihood of disrupting chromatid continuity may increase because of increased exposure of SSBs. Moreover, new SSBs might arise from the nicking activity of nucleases acting on single-stranded DNA.

When replication forks encounter polymerase-blocking lesions such as bulky adducts, DSBs can arise by several processes. A blocking lesion in the leading strand can result in the generation of extensive downstream ssDNA in that strand (Fig. 7A). These single-stranded gaps have been documented experimentally during SV40 replication that initiates upstream of a pyrimidine-dimer[173,174] and are favored under conditions when damage bypass of specific lesions is inhibited.[144,175] Under these conditions, ssDNA is generated as the replication fork proceeds while chain extension on the leading strand is restricted. The resultant DNA structures can become destabilized and cause the marked increase in DSBs seen after UV irradiation of both yeast[176] and hamster cells[177] containing photolytic lesions induced by UVA in BrdUrd-substituted DNA in the presence of a photosensitizing dye. This phenomenon is greatly exaggerated in bypass-deficient xeroderma pigmentosum variant (XP-V) cells, which are defective in the bypass polymerase Y/η (discussed further below).[178,179] Regions of ssDNA are also candidates for the formation of secondary structures such as hairpins and cruciform structures that can be recognized and cleaved by the MRN complex or other enzymes.[180-183]

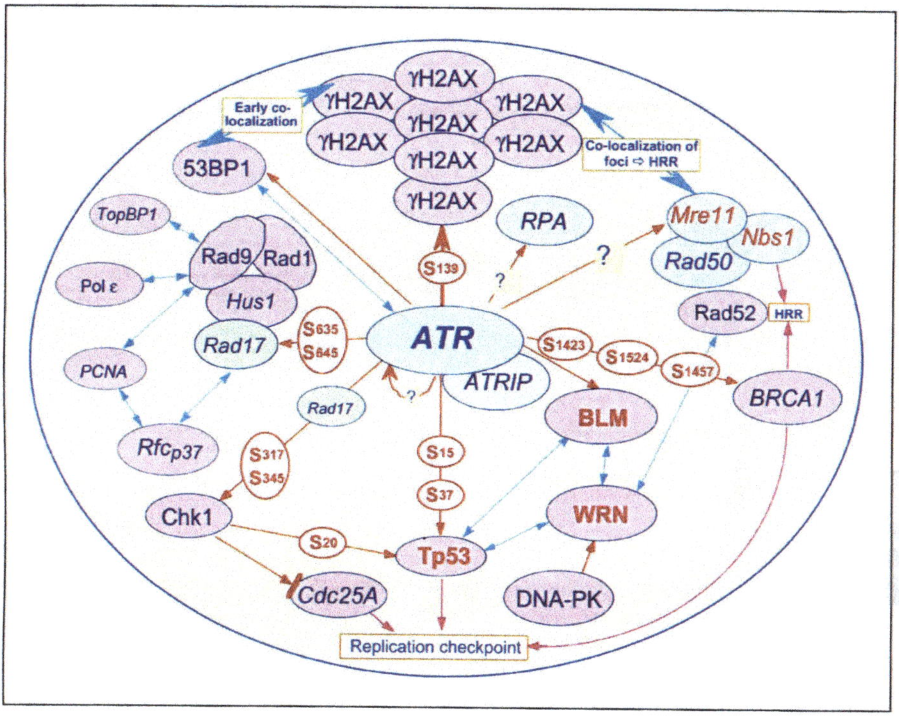

Figure 6. ATR-mediated phosphorylation events and interactions that implement signaling, repair, and checkpoint functions in response to replication-associated DSBs. ATR has diverse substrates that overlap with those of ATM. Proteins with names in red are involved in human genetic disorders and proteins in italics are required for viability of dividing cells. Site-specific ATR mediated phosphorylation events have been determined for H2AX,[251] ATRIP (ATR interacting protein),[210] BRCA1,[221,222] Tp53,[218,219,341] Chk1,[214,246,342] and Rad17.[246,343-345] ATR additionally phosphorylates BLM,[292] and possibly RPA in response to replication-associated DSBs (K. Cimprich, personal comm.). WRN is phosphorylated by DNA-PK, which inhibits its activity.[346] The interactions marked by blue arrows have been reported for ATR-53BP1,[90] BLM-Tp53,[302,347,348] BLM-WRN,[349] WRN-Tp53,[302,350] WRN-Rad52,[351] Rad9-TopBP1,[126] Rad9-Polε,[126] and (Rad9-PCNA, PCNA-RFCp37, RFCp37-Rad17).[352] Kinase Chk1 contributes to S and G2 checkpoints by phosphorylating and inactivating a key checkpoint phosphatase, Cdc25A.[353-355]

Under certain circumstances stalled replication forks can regress or reverse to generate a 4-stranded structure (often called a "chickenfoot"[184]) that can be considered a topologically masked DSB (Fig. 7E). Recent electron micrographs reveal the presence of such structures in yeast *rad53* kinase checkpoint mutants.[185] However, in mammalian cells early studies cast doubt on the existence of these structures. The appearance of doubly-dense DNA from cells pulse labeled with BrdUrd was found to decrease dramatically upon crosslinking with psoralen, suggesting that branch migration occurred as an artifact of DNA isolation.[186,187] Regressed forks are structurally similar to, but topologically distinct from, Holliday junctions (HJs), a common intermediate produced during the repair of DSBs by HR (Fig. 7C). These 4-stranded structures generated by fork regression can be unwound by the structure specific helicases WRN and BLM[188-190] and cut by resolvases.[191,192] Thus, branched intermediate structures that arise at sites of stalled replication can be converted to DSBs in a manner dependent upon their structural and topological context, as further addressed below.

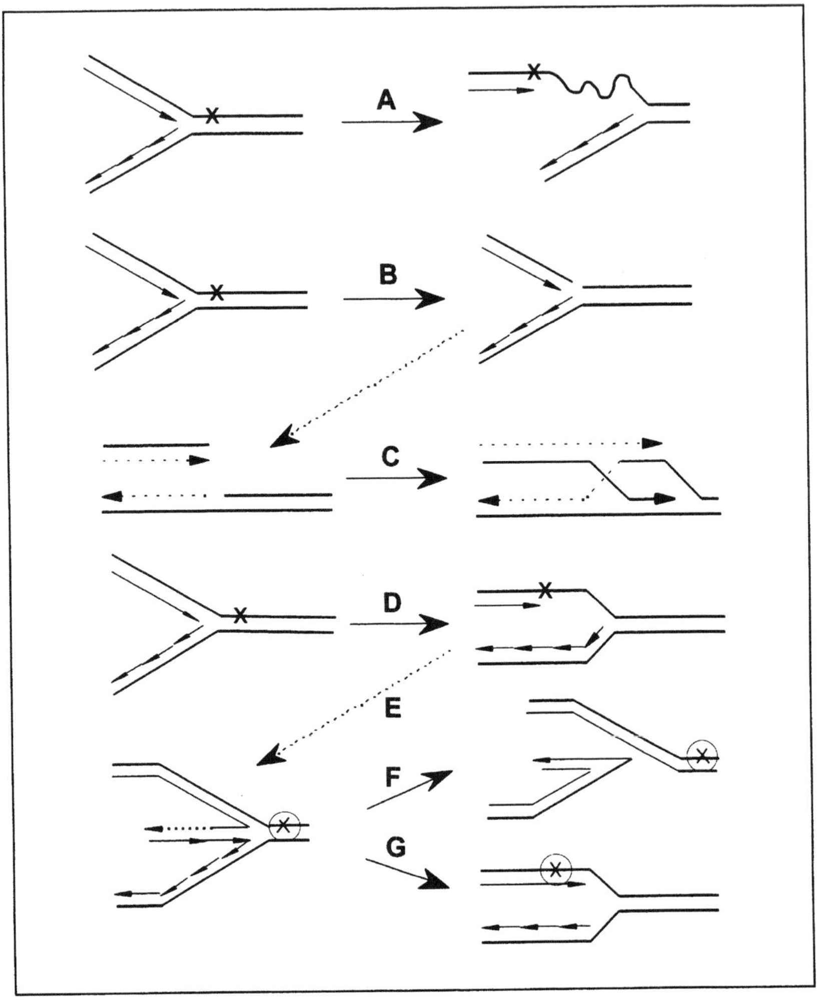

Figure 7. Formation and repair of DSB arising when replication forks encounter damage. (A) Arrest of the replication fork. (B) Fork collapse and DSB generation upon the encounter of the replication fork with a preexisting nick or gap. (C) Formation of a Holliday junction (HJ) at the site of a broken fork. (D) Replication fork encounters a leading strand lesion, designated "X", resulting in the generation of single-stranded DNA. (E) Fork regression (reversal) by branch migration generates a four-stranded chicken-foot structure; extension of the leading strand may occur in the sequence corresponding to that of the site of the lesion. (F) Resolution of the HJ by resolvase, resulting in an incompletely replicated chromosome. (G) Branch migration resets the replication fork with the lesion either present or absent.

Replication-Associated DSBs Arising from Inhibition of Replication

As in *E. coli* and yeast,[193-195] stalled replication forks caused by DNA synthesis inhibitors in mammalian cells are an efficient source of DSBs, cytotoxicity, and potentially deleterious recombination.[196] Inhibition of DNA replication by hydroxyurea and aphidicolin produces chromatid discontinuities at replication forks.[197,198] DSB-promoted intrachromosomal recombination of a partially duplicated *HPRT* gene in hamster cells was most strongly induced by inhibitors of DNA synthesis (hydroxyurea, methotrexate, aphidicolin, cytosine arabinoside), followed by the topoisomerase inhibitors (camptothecin, etoposide), then bifunctional alkylating agents (cisplatin, mitomycin C), and lastly spindle poisons (vincristine).[199] Camptothecin-induced DSBs associated with replication forks in synchronized cells were detectable by PFGE and were repaired primarily by HRR, not NHEJ.[199]

Although both HRR and NHEJ mediate the resolution of stalled forks during replication arrest, analysis of mutant phenotypes suggests the two pathways play different roles, which likely depend on the nature of fork damage.[197] There is some evidence that blocked forks that have not collapsed promote HRR,[197] but this issue needs further study. Related studies found a differential involvement of NHEJ and HRR in the resolution of DSBs accumulated during extended (6-24 hr) arrest induced by replication inhibitors.[200] The data suggest that NHEJ precedes HRR in the resolution of arrest-induced DSBs, and that HRR becomes more important with extended replication arrest induced by inhibitors.[200,201]

Recent studies in permeabilized cells treated with cisplatin provide insights into how chromosomal fork progression is controlled after DNA damage in vertebrates.[202] Cisplatin DNA adducts *actively* slow fork progression by a process that requires XRCC3, a Rad51 paralog that participates in homologous recombination.[203] The addition of purified human RAD51C-XRCC3 protein complex restores slowing of fork progression in permeabilized XRCC3$^{-/-}$ cells.[202] Moreover, this requirement for XRCC3 is alleviated by adding human Rad51 protein, but not by Rad52 or the recombination complex Rad51B-Rad51C-Rad51D-XRCC2. These data demonstrate that XRCC3 and Rad51 cooperatively modulate the progression of replication forks on damaged vertebrate chromosomes.

Recent results suggest that the mode of DSB repair at arrested replication forks depends on the functional status of Tp53 (see also discussion on WRN and BLM below). The induction of HRR during the inhibition of replication is stimulated by mutant Tp53,[204] and stalled replication forks produced by hydroxyurea and aphidicolin stimulate the accumulation of nuclear Tp53 that is impaired and altered in its transcriptional activity compared with Tp53 induced by IR.[205] In XP-V cells following UV-induced replication arrest, inactivation of Tp53 by SV40 or HPV16 transformation stimulates DSB-dependent modification and relocalization of the MRN complex.[178,179,206] High wortmannin concentrations, which inhibit PI3KKs (phosphatidylinositol-3-kinase related kinases), lead to a marked increase in the number of transformed XP-V cells exhibiting MRN foci following UV irradiation.[207]

Recognition and Signaling of Stalled and Collapsed Forks

ATR-Mediated Signaling

Creation of DSBs at sites of arrested forks can occur when enzymatic processing of the blocked structures leads to breakage of parental DNA strands, causing what is termed fork collapse (Fig. 7B). The details of how mammalian cells recognize and signal the presence of specific replication anomalies are beginning to be understood. Although the abnormal DNA structures that are initially recognized as critical replication fork "damage" are not yet well defined, the ATR kinase is a key early component in the signaling process.[3,38,169] (The homologs of ATR in yeasts are Rad3^{Sp} and Mec1^{Sc}.) It is not entirely clear whether replication-associated γH2AX formation *always* requires collapse of a replication fork. However, the fact that a 3 hr aphidicolin treatment did not produce γH2AX foci strongly suggests that the signal arises from DSBs rather than stalled forks.[53] (Much longer aphidicolin exposure causes γH2AX focus formation).[134] In case of treatment with camptothecin, which binds Top1, DSBs associated with the cleavage complex at replication forks result in ATR-dependent γH2AX focus formation; DNA-PK and ATM also contributes to this phosphorylation.[53] Examples of γH2AX foci seen in both CHO and human cells are given in (Fig. 3), panels B-D.

Even though ATR and ATM are functionally related, they differ very significantly. In contrast to ATM, which responds primarily to DSBs occurring outside the context of replication arrest, the ATR kinase prevents the accumulation of DSBs during replication arrest, as from aphidicolin treatment (Fig. 5).[3,169,208] Unlike ATM, deletion of ATR function leads to early mouse embryonic and cellular lethality.[147,209,210] ATR, but not ATM, plays a critical role in preventing chromosomal gaps and breaks at fragile sites (as well as at random sites), which are greatly enhanced by inhibiting replication with aphidicolin.[198]

The response of ATR to replication stress has provided many of the insights into its essential cellular functions as a regulator of replication arrest. Overexpression of kinase-inactive ATR mutants in human fibroblasts produces dominant negative phenotypes that show abnormalities in cell cycle progression, reduced phosphorylation of signaling and repair proteins, and elevated sensitivity to killing by DNA damaging agents (IR, UV, MMS, the topoisomerase I inhibitor topotecan, & the topoisomerase II inhibitor etoposide) and replication blocks (e.g., hydroxyurea).[42,211,212] In Xenopus egg extracts and mammalian cells, ATR is required both to phosphorylate Chk1 in response to replication arrest and to properly activate the replication checkpoint.[213-215] In the Xenopus extracts, replication forks were shown to be an obligate intermediate for the activation of this checkpoint.[216,217] In addition to phosphorylating Chk1, ATR plays a role in regulating the replication checkpoint through phosphorylation of Ser15 on Tp53, which is also a target of the Chk1 kinase (Fig. 5).[3,214,215,218-220] ATR and p53 can function independently, but loss of both can cause synergistic disruption of the replication checkpoint.[219] In response to replication stress induced by UV-dimer photoproducts or replication inhibitors, ATR redistributes to form nuclear foci, presumably at stalled or broken replication forks.[90,221,222] ATR interacts with 53BP1, phosphorylates it in vitro, and colocalizes with it after replication inhibition.[90] ATR also phosphorylates BRCA1 with an overlapping spectrum of sites compared with ATM, and forms foci that partially colocalize with BRCA1 foci.[221,222]

ATR appears to act, not only in response to drug-imposed blocks to replication (e.g., aphidicolin, hydroxyurea), but also as an intrinsic replication fork checkpoint initiator that is normally active throughout S phase.[222] Experiments using Xenopus extracts have shown that this checkpoint function depends on the replication-dependent chromatin-binding properties of ATR.[223] After the initiation of replication, ATR binds to chromatin, where it can phosphorylate a range of downstream effectors and then dissociate upon the completion of replication.[216,223] Nuclei in Xenopus extracts treated with replication inhibitors accumulate ssDNA, RPA, and γH2AX foci, consistent with the production of replication-associated DSBs.[224] In the absence of inhibitors, replication-associated RPA coating ssDNA appears to facilitate the binding to chromatin of both ATR and the checkpoint protein Hus1 (hydroxyurea sensitive); the recruitment of Polα is also required for chromatin association of Hus1.[172] Recruitment of the Hus1 protein complex to chromatin is independent of ATR binding, and checkpoint activation requires RNA synthesis by Polα.[172,225] Since both ATR and Hus1 are required for the phosphorylation/activation of Chk1,[172] their coincident binding to chromatin is likely critical in activating ATR either by interaction with Hus1 or by recruiting Chk1 to the chromatin.

Improper execution of the replication checkpoint may account for the chromosomal fragmentation and early embryonic lethality observed in ATR null mouse embryos[147,209] and in Chk1 null mice.[226] Somewhat surprisingly, incomplete DNA replication in mouse cells after aphidicolin treatment can prevent M-phase entry independently of ATR and inhibitory phosphorylation of Cdc2.[169] However, when the replication inhibitor is removed, ATR knockout cells proceed to mitosis with extensive chromosome breaks, indicating that ATR provides a key genome maintenance function in S phase.[169]

ATR may act directly as a DNA damage sensor during replication arrest. ATR preferentially binds to UV-damaged DNA in vitro.[227] This binding depends on UV fluence and full-length ATR, and results in a stimulation of its kinase activity. The binding partner of ATR, ATRIP (hRad26), was recently identified as the human homolog of $Ddc2^{Sc}$ (also called Lcd1 or Pie1) and $Rad26^{Sp}$.[210] Observations that ATRIP associates with ATR, is a substrate of ATR, and is a phosphoprotein in vivo,[210] are compatible with similar interactions found for the yeast homologs Mec1-Ddc2 in *S. cerevisiae* and Rad3-Rad26 in *S. pombe*.[228-231] DNA damage and replication inhibition cause ATRIP to colocalize with ATR.[210] Deletion of ATRIP reduces the level of the ATR protein and generates checkpoint defects similar to those of an ATR deletion,[210] which suggests that ATRIP and ATR are mutually dependent in signaling and checkpoint pathways. It remains unclear whether ATRIP recruits and/or stimulates the association of ATR to replication-associated DSBs.

The Rad17 and Rad9 Complexes

The ATR/ATRIP complex interacts with other early damage sensor elements during checkpoint activation. Specifically, the mammalian checkpoint proteins Rad17 (RFC1 homolog; RFC = replication factor C)[232-234] and the Rad9-Rad1-Hus1 (9-1-1) complex[235-237] are named after their counterparts in *S. pombe*.[4] (The corresponding proteins in *S. cerevisiae* are Rad24, Rad17, Ddc1, and Mec3, respectively.) While direct evidence is lacking that the Rad17-RFC heteropentameric complex and the 9-1-1 complex actually recognize replication-associated DSBs or other abnormal DNA structures at arrested forks, their interactions with chromatin and ATR appear to be critical in the early recognition and signaling of replication arrest. The Rad17-RFC and 9-1-1 complexes show sequence and structural similarity to the clamp loader (RFC) and sliding clamp (PCNA) complexes required for replication.[216,232,238-243] The biological importance of Rad9 in IR sensitivity is illustrated by the strong phenotype of *rad9* knockout mouse ES cells, which display increased spontaneous chromosomal aberrations, IR and UV sensitivity to killing (3-fold), and a partially defective G2 checkpoint.[250] Homozygous mutant embryos die between E9.5 and E12.5.

It was recently inferred that the checkpoint role of the 9-1-1 complex is not restricted to S phase and replication blockage. DNA-damage-induced binding of Rad9 to chromatin occurred in noncycling cells after exposure to IR or a bulky-adduct mutagen.[244] However, the dose of IR used (50 Gy) may produce interactions that are irrelevant to normal physiological responses. In cycling cell populations, the IR- and hydroxyurea-induced binding of Rad9 (and Rad1) to chromatin occurs independently of the ATM phosphorylation of Rad9 at Ser^{272} (Fig. 2) and PIKK activities. Phosphorylation is also not required for Rad9's interaction with Rad1, Hus1, and Rad17.[244]

The critical role of Rad17 in preventing accumulation of DNA DSBs during replication is revealed by the properties of a *Rad17flox/-* conditional mutant in human HCT116 cells.[245] Loss of Rad17 causes rapid accumulation of chromosomal breaks and rearrangements as well as endoreduplication. However, the chromosomal breakage in *rad17* null cells was less severe than that of *atr* null cells examined in parallel.[245] Rad17 null cells have defective Chk1 phosphorylation after UV damage, normal phosphorylation of ATM targets including Chk2 after IR damage, and a partially defective G2 checkpoint after IR damage.

Rad17 is constitutively bound to chromatin in human cells (although this was not seen in Xenopus egg extracts without DNA damage[217]) and is phosphorylated by ATR on chromatin after treatment with UV radiation, γ-rays, and hydroxyurea.[246] Rad17, but not its phosphorylation by ATR, is required for loading the 9-1-1 complex onto chromatin (Fig. 6) in a manner analogous to PCNA loading by RFC. However, phosphorylation of Rad17 is required for the downstream phosphorylation/activation of Chk1. Both ATM and ATR are required for Rad17 phosphorylation in response to IR at early times, but UV-induced Rad17 phosphorylation appears to be specifically produced by ATR.[246] Hydroxyurea-induced Rad17 phosphorylation is partially dependent on ATR but independent of ATM in the presence of ATR.[246] Similar to

Xenopus mentioned above, the 9-1-1 and ATR-ATRIP complexes in human cells can be recruited to chromatin independently (in response to UV damage). Both complexes are present at sites of UV damage as indicated by the partial colocalization of ATR foci with Rad17-Ser^{635} foci. Thus, Rad17's interactions with the 9-1-1 complex may help determine the selection of substrates available to ATR.[246] In *S. pombe* recent evidence indicates that the DinB damage-response polymerase physically interacts with the 9-1-1 complex and requires Rad17 to associate with chromatin, suggesting that the checkpoint response includes translesion synthesis.[247] One model of checkpoint activation is that the loading and interaction of the ATR-ATRIP and 9-1-1 complexes, mediated by Rad17, creates a higher-order chromatin structure to facilitate signaling and phosphorylation of Chk1.[246] Such a chromatin change might be analogous to that which appears to be mediated by ATM already discussed.[48]

This idea is compatible with the finding that Hus1 acts upstream of Chk1 and is required for its optimal phosphorylation in mammalian cells.[248] Hus1 is not required for Tp53 accumulation and activation or for Chk2 phosphorylation. Disrupted signaling during replication stress likely underlies the embryonic lethality in Hus1-deficient mice.[249] Hus1-deficient embryonic fibroblasts have been rescued for in vitro viability by simultaneous disruption of CDKN1A/p21,[249] and these cells exhibit chromosomal instability, heightened sensitivity to replication blocks, and altered cell cycle responses. They display high sensitivity to UV radiation and hydroxyurea but only slight IR sensitivity,[248,249] which is consistent with the idea that the 9-1-1 complex functions in the response to replication-associated damage. The lesser IR sensitivity of *hus1* mutant cells[248] compared with *rad9* cells[250] suggests that these two complex members have overlapping but not identical functions.

Factors Promoting the Repair of Replication Associated DSBs

As already discussed, overt DSB production leads to the production of γH2AX, a large-scale chromatin modification that can be visualized as nuclear foci. Exposure of mammalian cells to UV radiation or hydroxyurea leads to ATR-dependent phosphorylation and γH2AX focus formation.[251] Colocalization of these γH2AX foci with PCNA in S-phase synchronized cultures suggests that these foci are associated with sites of replication fork arrest.[251] The idea that these γH2AX foci reflect DSBs is supported by the finding that both UV radiation and hydroxyurea efficiently induce sister-chromatid exchange,[252] a manifestation of DSB repair by homologous recombination involving crossing-over.[253] Replication-dependent formation of γH2AX foci is also seen in both human cells [53] and nuclei incubated in Xenopus egg extracts [224] after treatment with the topoisomerase I poison camptothecin. In the Xenopus system, the induction of γH2AX foci is inhibited by geminin, a replication licensing inhibitor.

Additional evidence supporting the importance of γH2AX after S-phase DSB formation comes from studies utilizing replication-defective xeroderma pigmentosum variant (XP-V) cells. Exposure of XP-V cells to UV radiation leads to a fluence-dependent increase in the fraction of cells showing γH2AX foci, which are illustrated in (Fig. 3) These foci are only observed in S-phase cells, and they coincide with PCNA and MRN foci, further supporting the concept that the chromatin modification associated with H2AX phosphorylation recruits HRR proteins that facilitate repair between sister chromatids.[178,179] Moreover, cells derived from H2AX null mice show a decrease in the portion of proliferating cells and an increased level of S-phase-derived chromatid aberrations.[76] These observations, coupled with data showing that $H2AX^{\Delta/\Delta}$ ES cells exhibit reduced HRR[76] but normal NHEJ,[77] suggest that chromatin modification involving H2AX is critical in protecting cells against replication-associated genomic instability associated with aberrant recombination.

The Mre11 complex plays an important role in repairing replication associated DSBs. MRN exhibits extensive colocalization with PCNA throughout S phase, and replication fork stalling imposed by hydroxyurea enhances the chromatin association of MRN.[114,254] As in

XP-V cells discussed above, these results further suggest that MRN loads onto chromatin at blocked or collapsed replication forks. This idea is supported by the observation that MRN preferentially localizes to single-stranded DNA arising in hydroxyurea-treated cells.[254] In camptothecin treated cells, MRN focus formation requires γH2AX formation, and H2AX null mouse cells are hypersensitive to killing by camptothecin.[53] The finding that aphidicolin blocks camptothecin-induced γH2AX focus formation shows that replication produces these DSBs.[53]

TopBP1 appears to be another key protein involved in preventing replication-associated chromosomal rearrangements. TopBP1contains eight BRCT domains, which mediate multiple interactions, and has sequence homologs in yeast[255,256] and flies.[257] These homologs play important roles in DNA replication, repair and checkpoints in lower organisms.[257-261] In addition to facilitating normal replication through its interaction with Pol-ε, TopBP1 responds to the inhibition of DNA synthesis by localizing with other repair proteins (BRCA1, PCNA) during S-phase, suggesting a possible role in rescuing stalled forks.[126,127] TopBP1 localized at sites of replication arrest may act to relieve torsional stress developed during the generation of anomalous DNA structures. Replication stress also elicits focus formation for 53BP1,[91] a protein discussed earlier in the context of IR damage.

Processing Abnormal Replication Intermediates and Associated DSBs

Helicases and topoisomerases are specialized enzymes that modify the three dimensional structure of DNA. Helicases increase accessibility of the replication and repair machinery to DNA by locally unwinding the duplex. Topoisomerases modulate the torsional strain of the DNA helix by catalyzing the interconversion of topological isomers. Known interactions between these two classes of proteins suggest a need to colocalize their activities to resolve abnormal DNA structures that can arise at stalled forks.

The RecQ family of helicases is critical to the maintenance of genomic integrity. These helicases (see reviews in refs. 190,262,263), named after the *E. coli recQ* gene product, include yeast Sgs1[Sc264,265] and Rqh1[Sp,266,267] and five members in humans: BLM,[268] WRN,[269] RecQ1/RecQL,[270,271] RecQ4,[272] and RecQ5.[272-274] Deficiencies in the BLM, WRN, and RecQ4 helicases cause Bloom's and Werner's syndromes,[268,269] and some cases of Rothmund-Thomson syndrome.[275-277,279] These rare genetic diseases manifest distinct yet overlapping clinical phenotypes of immunodeficiency, premature aging, chromosomal instability, and predisposition to cancer (see reviews in refs. 25,189,190,280-282).

The mutations in BLM and WRN helicases are associated with replication defects, including impaired progression of replication forks, an accumulation of abnormal replication intermediates,[283-285] and aberrant homologous recombination.[286,287] BLM- and WRN-defective cells display an abnormally high percentage of deletion mutations at specific loci.[288,289] BLM and WRN helicases can suppress the increased homologous and illegitimate recombination in the *S. cerevisiae sgs1* mutant.[290] Elevated sensitivity to replication-blocking inhibitors is seen in Bloom's syndrome (BS) cell lines[291-294] and in WRN-deficient cells.[295-299] Conflicting results are reported concerning altered sensitivity of BLM-deficient cells to inhibition by hydroxyurea.[291,292,294] WRN mutant cell lines consistently show hypersensitivity to camptothecin and defective responses to hydroxyurea.[295-299]

It has been proposed that DSBs formed during replication arrest can lead to the formation of HJs (see Fig. 7C), substrates recognized by the BLM and WRN helicases.[188,300,301] The ability of these helicases to unwind HJs is dependent on Tp53, which binds to the enzymes and attenuates their branch migration activity, and possibly their anti-recombinase functions.[302] Tp53 and BLM functionally interact during resolution of stalled DNA replication forks.[278] The evidence supports a model in which the disruption of abnormal structures by the BLM and WRN helicases prevents aberrant recombination events that would result in chromosomal rearrangement (see Fig. 7G).[188,287,300,301,303] The interaction between BLM and topoisomerase

IIIα (TopIII) in human cells, which is highly conserved across eukaryotic species,[264,304-306] may also promote unlinking of the parental duplex by TopIII at sites of paused or convergent replication forks.

Once formed, HJs may elude the activity of helicases and be cleaved by endonucleolytic HJ resolvasesto generate DSBs. Much of our understanding of these reactions comes from the genetic studies in bacterial systems characterizing the branch migration complex RuvA/RuvB, and the RuvC and RusA resolvases (see ref. 192 for a brief review). Activities from mammalian cell extracts that resemble those expected for a true HJ resolvase have been identified,[307,308] but it is unclear whether these activities correspond to the newly described mus81 proteins in yeast and humans.[191,309-311] Nonetheless, the Mus81 homologs appear to specifically cleave replication intermediates that possess branched structures, and Mus81 (i.e., the Mus81-Eme1 heterodimeric endonuclease) is particularly important in *S. pombe* in processing stalled or collapsed forks in a RecQ helicase-deficient background.[311]

Recent work shows an important link between the BLM helicase and the MRN complex during replication arrest. In response to hydroxyurea, the formation of MRN foci at sites of stalled forks was sharply reduced in Bloom's syndrome cells.[292] However, in nonBS cells, ATR-dependent phosphorylation of BLM was not required for subnuclear relocalization of MRN.[292] Function(s) of the MRN complex are suggested by its unique architecture.[72,139,140] The Rad50 zinc-hook motif, through the interaction of two Rad50 "tails", provides a means of tethering sister chromatids for recombinational repair, thereby limiting the undesirable dissociation of broken DNA ends. After fork arrest, the MRN complex may assist in repairing DSBs both by promoting inter-sister connectivity lost during fork regression and by helping to restore chromatid continuity during fork breakdown or HJ resolution. Notably, MRN foci that appear in S phase detergent-extracted cells colocalize with PCNA foci during DNA replication and appear normal in $H2AX^{\Delta/\Delta}$ mutant cells.[76]

Outlook

We discussed what is known about the events of DSB formation, recognition, and signaling, which facilitate the recruitment of checkpoint and repair proteins to these lesions. Homologous recombination may well have arisen early in life forms as a mechanism for ensuring chromosome continuity in the face of DSBs that arise normally during replication. The large genomes of higher eukaryotes must achieve remarkable accuracy in rapidly detecting each DSB and announcing its presence through signal amplification and transduction in order to keep the genome intact in each daughter cell at mitosis. The discovery of γH2AX foci as a likely bona fide marker and sentinel for DSBs represents a major advance.[11,49,50] It appears likely that γH2AX foci will provide a reliable method of quantifying DSBs and their repair.[11] Just how this modification is triggered through ATM's chromatin sensing capacity remains to be determined. Additional candidates for the initial sensors, which help mediate γH2AX formation, are under intense investigation.

Recent research brings the promise of providing the cytological tools needed to accurately quantify the levels of DSBs in fixed cells, and perhaps eventually in living cells.[73] Measuring the very early colocalization of 53BP1 and MDC1 with γH2AX may add robustness to experiments designed to accurately measure levels of DSBs at low IR doses. Better quantitative methods of image analysis are greatly needed to remove the subjectivity and labor inherent in the visual scoring of cytological foci. Since rapid progress has been made in recent years in this field, it seems very likely that we will soon have a detailed understanding of how cells cope so efficiently with DSBs.

Acknowledgments

We thank Karlene Cimprich, John Hinz, David Schild, Robert Tebbs, Alice Yamada, and Dawn Yean for providing valuable comments on the manuscript, and also Joel Bedford, Bill Dewey, and John Ward for helpful discussion. This work was prepared under the auspices of the US Department of Energy by Lawrence Livermore National Laboratory under contract No. W-7405-ENG-48 and funded by the Low Dose Radiation Research Program, Biological and Environmental Research (BER), U.S. Department of Energy, by NIH/NCI grant P01 CA92584-02, and by ACS grant # RPG-00-036-01 CNE (CLL).

References

1. Hoeijmakers JH. Genome maintenance mechanisms for preventing cancer. Nature 2001; 411:366-374.
2. Zhou BB, Elledge SJ. The DNA damage response: Putting checkpoints in perspective. Nature 2000; 408:433-439.
3. Abraham RT. Cell cycle checkpoint signaling through the ATM and ATR kinases. Genes Dev 2001; 15:2177-2196.
4. Melo J, Toczyski D. A unified view of the DNA-damage checkpoint. Curr Opin. Cell Biol 2002; 14:237-245.
5. Ruiz de Almodovar JM, Steel GG, Whitaker SJ et al. A comparison of methods for calculating DNA double-strand break induction frequency in mammalian cells by pulsed-field gel electrophoresis. Int J Radiat Biol 1994; 65:641-649.
6. Cedervall B, Wong R, Albright N et al. Methods for the quantification of DNA double-strand breaks determined from the distribution of DNA fragment sizes measured by pulsed-field gel electrophoresis. Radiat Res 1995; 143:8-16.
7. Ward JF. DNA damage produced by ionizing radiation in mammalian cells: Identities, mechanisms of formation, and reparability. Prog Nucl Acids Res Mol Biol 1988; 35:95-125.
8. Ward JF. Nature of lesions formed by ionizing radiation, DNA damage and repair. In: Nickoloff JA, Hoekstra M, eds. DNA Repair in Higher Eukaryotes. Totowa: Humana Press, 1998:2:65-84.
9. Rogakou EP, Nieves-Neira W, Boon C et al. Initiation of DNA fragmentation during apoptosis induces phosphorylation of H2AX histone at serine 139. J Biol Chem 2000; 275:9390-9395.
10. Sedelnikova OA, Rogakou EP, Panyutin IG et al. Quantitative detection of (125)IdU-induced DNA double-strand breaks with gamma-H2AX antibody. Radiat Res 2002; 158:486-492.
11. Rothkamm K, Lobrich M. Evidence for a lack of DNA double-stranded break repair in himan cells exposed to very low x-ray doses. Proc Natl Acad Sci USA 2003; 100:5057-5062.
12. Cornforth MN, Bedford JS. On the nature of a defect in cells from individuals with ataxia-telangiectasia. Science 1985; 227:1589-1591.
13. Girard PM, Foray N, Stumm M et al. Radiosensitivity in Nijmegen Breakage Syndrome cells is attributable to a repair defect and not cell cycle checkpoint defects. Cancer Res 2000; 60:4881-4888.
14. Wang H, Zeng ZC, Bui TA et al. Efficient rejoining of radiation-induced DNA double-strand breaks in vertebrate cells deficient in genes of the RAD52 epistasis group. Oncogene 2001; 20:2212-2224.
15. Beckman KB, Saljoughi S, Mashiyama ST et al. A simpler, more robust method for the analysis of 8-oxoguanine in DNA. Free radic. Biol Med 2000; 29:357-367.
16. Beckman KB, Ames BN. Oxidative decay of DNA. J Biol. Chem 1997; 272:19633-19636.
17. Sutherland BM, Bennett PV, Sidorkina O et al. Clustered damages and total lesions induced in DNA by ionizing radiation: Oxidized bases and strand breaks. Biochemistry 2000; 39:8026-8031.
18. Sutherland BM, Bennett PV, Sidorkina O et al. Clustered DNA damages induced in isolated DNA and in human cells by low doses of ionizing radiation. Proc Natl Acad Sci USA 2000; 97:103-108.
19. Sutherland BM, Bennett PV, Sutherland JC et al. Clustered DNA damages induced by x rays in human cells. Radiat Res 2002; 157:611-616.
20. Dianov GL, O'Neill P, Goodhead DT. Securing genome stability by orchestrating DNA repair: Removal of radiation-induced clustered lesions in DNA. Bioessays 2001; 23:745-749.

21. Barnes DE. Nonhomologous end joining as a mechanism of DNA repair. Curr Biol 2001; 11:R455-R457.
22. Pierce AJ, Jasin M. NHEJ deficiency and disease. Mol Cell 2001; 8:1160-1161.
23. Thompson LH, Schild D. The contribution of homologous recombination in preserving genome integrity in mammalian cells. Biochimie 1999; 81:87-105.
24. Thompson LH, Schild D. Homologous recombinational repair of DNA ensures mammalian chromosome stability. Mutat Res 2001; 477:131-153.
25. Thompson LH, Schild D. Recombinational DNA repair and human disease. Mutat Res 2002; 509:49-78.
26. Karanjawala ZE, Grawunder U, Hsieh CL et al. The nonhomologous DNA end joining pathway is important for chromosome stability in primary fibroblasts. Curr Biol 1999; 9:1501-1504.
27. Difilippantonio MJ, Zhu J, Chen HT et al. DNA repair protein Ku80 suppresses chromosomal aberrations and malignant transformation. Nature 2000; 404:510-514.
28. Ferguson DO, Sekiguchi JM, Chang S et al. The nonhomologous end joining pathway of DNA repair is required for genomic stability and the suppression of translocations. Proc Natl Acad Sci USA 2000; 97:6630-6633.
29. Karanjawala ZE, Murphy N, Hinton DR et al. Oxygen metabolism causes chromosome breaks and is associated with the neuronal apoptosis observed in DNA double-strand break repair mutants. Curr Biol 2002; 12:397-402.
30. Barnes DE, Stamp G, Rosewell I et al. Targeted disruption of the gene encoding DNA ligase IV leads to lethality in embryonic mice. Curr Biol 1998; 8:1395-1398.
31. Gao Y, Sun Y, Frank KM et al. A critical role for DNA end-joining proteins in both lymphogenesis and neurogenesis. Cell 1998; 95:891-902.
32. Li GC, Ouyang H, Li X et al. Ku70: A candidate tumor suppressor gene for murine T cell lymphoma. Mol Cell 1998; 2:1-8.
33. Vogel H, Lim DS, Karsenty G et al. Deletion of Ku86 causes early onset of senescence in mice. Proc Natl Acad Sci USA 1999; 96:10770-10775.
34. Gu Y, Sekiguchi J, Gao Y et al. Defective embryonic neurogenesis in Ku-deficient but not DNA-dependent protein kinase catalytic subunit-deficient mice. Proc Natl Acad Sci USA 2000; 97:2668-2673.
35. Kemp LM, Jeggo PA. Radiation-induced chromosome damage in X-ray-sensitive mutants (xrs) of the Chinese hamster ovary cell line. Mutat Res 1986; 166:255-263.
36. Darroudi F, Natarajan AT. Cytological characterization of Chinese hamster ovary X-ray-sensitive mutant cells xrs 5 and xrs 6. I. Induction of chromosomal aberrations by X-irradiation and its modulation with 3-aminobenzamide and caffeine. Mutat Res 1987; 177:133-148.
37. Takata M, Sasaki MS, Sonoda E et al. Homologous recombination and nonhomologous end joining pathways of DNA double-strand break repair have overlapping roles in the maintenance of chromosomal integrity in vertebrate cells. EMBO J 1998; 17:5497-5508.
38. Durocher D, Jackson SP. DNA-PK, ATM and ATR as sensors of DNA damage: Variations on a theme? Curr Opin Cell Biol 2001; 13:225-231.
39. Khanna KK, Lavin MF, Jackson SP et al. ATM, a central controller of cellular responses to DNA damage. Cell Death Differ 2001; 8:1052-1065.
40. Shiloh Y, Kastan MB. ATM: Genome stability, neuronal development, and cancer cross paths. Adv Cancer Res 2001; 83:209-254.
41. Shiloh Y. ATM and related protein kinases: Safeguarding genome integrity. Nat. Rev. Cancer 2003; 3:155-168.
42. Wright JA, Keegan KS, Herendeen DR et al. Protein kinase mutants of human ATR increase sensitivity to UV and ionizing radiation and abrogate cell cycle checkpoint control. Proc Natl Acad Sci USA 1998; 95:7445-7450.
43. Shiloh Y. ATM and ATR: Networking cellular responses to DNA damage. Curr Opin Genet Dev 2001; 11:71-77.
44. Wang Y, Cortez D, Yazdi P et al. BASC, a super complex of BRCA1-associated proteins involved in the recognition and repair of aberrant DNA structures. Genes Dev 2000; 14:927-939.
45. Banin S, Moyal L, Shieh S et al. Enhanced phosphorylation of p53 by ATM in response to DNA damage. Science 1998; 281:1674-1677.

46. Canman CE, Lim DS, Cimprich KA et al. Activation of the ATM kinase by ionizing radiation and phosphorylation of p53. Science 1998; 281:1677-1679.
47. Kozlov S, Gueven N, Keating K et al. ATP activates ataxia-telangiectasia mutated (ATM) in vitro. Importance of autophosphorylation. J Biol Chem 2003; 278:9309-9317.
48. Bakkenist CJ, Kastan MB. DNA damage activates ATM through intermolecular autophosphorylation and dimer dissociation. Nature 2003; 421:499-506.
49. Rogakou EP, Pilch DR, Orr AH et al. DNA double-stranded breaks induce histone H2AX phosphorylation on serine 139. J Biol Chem 1998; 273:5858-5868.
50. Rogakou EP, Boon C, Redon C et al. Megabase chromatin domains involved in DNA double-strand breaks in vivo. J Cell Biol 1999; 146:905-916.
51. Burma S, Chen BP, Murphy M et al. ATM phosphorylates histone H2AX in response to DNA double-strand breaks. J Biol Chem 2001; 276:42462-42467.
52. Kobayashi J, Tauchi H, Sakamoto S et al. NBS1 localizes to gamma-H2AX foci through interaction with the FHA/BRCT domain. Curr Biol 2002; 12:1846-1851.
53. Furuta T, Takemura H, Liao ZY et al. Phosphorylation of histone H2AX and activation of Mre11, Rad50, and Nbs1 in response to replication-dependent DNA double-strand breaks Induced by mammalian DNA topoisomerase I cleavage complexes. J Biol Chem 2003; 278:20303-20312.
54. Smith GC, Cary RB, Lakin ND et al. Purification and DNA binding properties of the ataxia-telangiectasia gene product ATM. Proc Natl Acad Sci USA 1999; 96:11134-11139.
55. Adamson AW, Kim WJ, Shangary S et al. ATM is activated in response to MNNG-induced DNA alkylation. J Biol Chem 2002; 277:38222-38229.
56. Andegeko Y, Moyal L, Mitelman L et al. Nuclear retention of ATM at sites of DNA double strand breaks. J Biol Chem 2001; 276:38224-38230.
57. Wang B, Matsuoka S, Carpenter PB et al. 53BP1, a mediator of the DNA damage checkpoint. Science 2002; 298:1435-1438.
58. Stewart GS, Wang B, Bignell CR et al. MDC1 is a mediator of the mammalian DNA damage checkpoint. Nature 2003; 421:961-966.
59. Scott SP, Bendix R, Chen P et al. Missense mutations but not allelic variants alter the function of ATM by dominant interference in patients with breast cancer. Proc Natl Acad Sci USA 2002; 99:925-930.
60. Spring K, Ahangari F, Scott SP et al. Mice heterozygous for mutation in Atm, the gene involved in ataxia-telangiectasia, have heightened susceptibility to cancer. Nat. Genet 2002; 32:185-190.
61. Lim DS, Kim ST, Xu B et al. ATM phosphorylates p95/nbs1 in an S-phase checkpoint pathway. Nature 2000; 404:613-617.
62. Wu X, Ranganathan V, Weisman DS et al. ATM phosphorylation of Nijmegen breakage syndrome protein is required in a DNA damage response. Nature 2000; 405:477-482.
63. Gatei M, Young D, Cerosaletti KM et al. ATM-dependent phosphorylation of nibrin in response to radiation exposure. Nat Genet 2000; 25:115-119.
64. Zhao S, Weng YC, Yuan SS et al. Functional link between ataxia-telangiectasia and Nijmegen breakage syndrome gene products. Nature 2000; 405:473-477.
65. Dong Z, Zhong Q, Chen PL. The Nijmegen breakage syndrome protein is essential for Mre11 phosphorylation upon DNA damage. J Biol Chem 1999; 274:19513-19516.
66. Shiloh Y. ATM: A sentry at the gate of genome instability. The Univ of Texas M D Anderson Cancer Center 55th Ann Symp Fund Cancer Res. Houston, TX: Maintenance of genomic integrity Oct 15-18. 2002:35. Abstract.
67. Haaf T, Golub EI, Reddy G et al. Nuclear foci of mammalian Rad51 recombination protein in somatic cells after DNA damage and its localization in synaptonemal complexes. Proc Natl Acad Sci USA 1995; 92:2298-2302.
68. Ashley T, Plug AW, Xu J et al. Dynamic changes in Rad51 distribution on chromatin during meiosis in male and female vertebrates. Chromosoma 1995; 104:19-28.
69. Haaf T, Hayman DL, Schmid M. Quantitative determination of rDNA transcription units in vertebrate cells. Exp Cell Res 1991; 193:78-86.
70. Petrini JH. The Mre11 complex and ATM: Collaborating to navigate S phase. Curr Opin Cell Biol 2000; 12:293-296.

71. D'Amours D, Jackson SP. The Mre11 complex: At the crossroads of DNA repair and checkpoint signalling. Nat Rev Mol Cell Biol 2002; 3:317-327.
72. Connelly JC, Leach DR. Tethering on the brink: The evolutionarily conserved Mre11-Rad50 complex. Trends Biochem Sci 2002; 27:410-418.
73. Essers J, Houtsmuller AB, van Veelen L et al. Nuclear dynamics of RAD52 group homologous recombination proteins in response to DNA damage. EMBO J 2002; 21:2030-2037.
74. Li L, Sharipo A, Chaves-Olarte E et al. The Haemophilus ducreyi cytolethal distending toxin activates sensors of DNA damage and repair complexes in proliferating and nonproliferating cells. Cell Microbiol 2002; 4:87-99.
75. Paull TT, Rogakou EP, Yamazaki V et al. A critical role for histone H2AX in recruitment of repair factors to nuclear foci after DNA damage. Curr Biol 2000; 10:886-895.
76. Celeste A, Petersen S, Romanienko PJ et al. Genomic instability in mice lacking histone H2AX. Science 2002; 296:922-927.
77. Bassing CH, Chua KF, Sekiguchi J et al. Increased ionizing radiation sensitivity and genomic instability in the absence of histone H2AX. Proc Natl Acad Sci USA 2002; 99:8173-8178.
78. Fernandez-Capetillo O, Chen HT, Celeste A et al. DNA damage-induced G(2)-M checkpoint activation by histone H2AX and 53BP1. Nat Cell Biol 2002; 4:993-997.
79. Shang YL, Bodero AJ, Chen PL. NFBD1, a novel nuclear protein with signature motifs of FHA and BRCT, and an internal 41 amino acid repeat sequence, is an early participant in DNA damage response. J Biol Chem 2002; 278:6323-6329.
80. Lou Z, Minter-Dykhouse K, Wu X et al. MDC1 is coupled to activated CHK2 in mammalian DNA damage response pathways. Nature 2003; 421:957-961.
81. Goldberg M, Stucki M, Falck J et al. MDC1 is required for the intra-S-phase DNA damage checkpoint. Nature 2003; 421:952-956.
82. Peng A, Chen PL. NFBD1, like 53BP1, is an early and redundant transducer mediating Chk2 phosphorylation in response to DNA damage. J Biol Chem 2003; 278:8873-8876.
83. Weinert TA, Hartwell LH. The RAD9 gene controls the cell cycle response to DNA damage in Saccharomyces cerevisiae. Science 1988; 241:317-322.
84. Schiestl RH, Reynolds P, Prakash S et al. Cloning and sequence analysis of the Saccharomyces cerevisiae RAD9 gene and further evidence that its product is required for cell cycle arrest induced by DNA damage. Mol Cell Biol 1989; 9:1882-1896.
85. Callebaut I, Mornon JP. From BRCA1 to RAP1: A widespread BRCT module closely associated with DNA repair. FEBS Lett 1997; 400:25-30.
86. Iwabuchi K, Bartel PL, Li B et al. Two cellular proteins that bind to wild-type but not mutant p53. Proc Natl Acad Sci USA 1994; 91:6098-6102.
87. Iwabuchi K, Li B, Massa HF et al. Stimulation of p53-mediated transcriptional activation by the p53-binding proteins, 53BP1 and 53BP2. J Biol Chem 1998; 273:26061-26068.
88. Schultz LB, Chehab NH, Malikzay A et al. p53 binding protein 1 (53BP1) is an early participant in the cellular response to DNA double-strand breaks. J Cell Biol 2000; 151:1381-1390.
89. Ward IM, Minn K, Van Deursen J et al. p53 binding protein 53BP1 is required for DNA damage responses and tumor suppression in mice. Mol Cell Biol 2003; 23:2556-2563.
90. Morales JC, Xia Z, Lu T et al. Role for the BRCA1 C-terminal repeats (BRCT) protein 53BP1 in maintaining genomic stability. J Biol Chem 2003; 278:14971-14977.
91. Rappold I, Iwabuchi K, Date T et al. Tumor suppressor p53 binding protein 1 (53BP1) is involved in DNA damage-signaling pathways. J Cell Biol 2001; 153:613-620.
92. Ward IM, Minn K, Jorda KG et al. Accumulation of checkpoint protein 53BP1 at DNA breaks involves its binding to phosphorylated histone H2AX. J Biol Chem 2003.
93. Anderson L, Henderson C, Adachi Y. Phosphorylation and rapid relocalization of 53BP1 to nuclear foci upon DNA damage. Mol Cell Biol 2001; 21:1719-1729.
94. Jullien D, Vagnarelli P, Earnshaw WC et al. Kinetochore localisation of the DNA damage response component 53BP1 during mitosis. J Cell Sci 2002; 115:71-79.
95. Abraham RT. Checkpoint signalling: Focusing on 53BP1. Nat Cell Biol 2002; 4:E277-E279.
96. DiTullio RA, Mochan TA, Venere M et al. 53BP1 functions in an ATM-dependent checkpoint pathway that is constitutively activated in human cancer. Nat Cell Biol 2002; 4:998-1002.

97. Kao GD, McKenna WG, Guenther MG et al. Histone deacetylase 4 interacts with 53BP1 to mediate the DNA damage response. J Cell Biol 2003; 160:1017-1027.
98. Bartek J, Falck J, Lukas J. CHK2 kinase—a busy messenger. Nat Rev Mol Cell Biol 2001; 2:877-886.
99. McGowan CH. Checking in on Cds1 (Chk2): A checkpoint kinase and tumor suppressor. Bioessays 2002; 24:502-511.
100. Hirao A, Cheung A, Duncan G et al. Chk2 is a tumor suppressor that regulates apoptosis in both an ataxia telangiectasia mutated (ATM)-dependent and an ATM-independent manner. Mol Cell Biol 2002; 22:6521-6532.
101. Takai H, Naka K, Okada Y et al. Chk2-deficient mice exhibit radioresistance and defective p53-mediated transcription. EMBO J 2002; 21:5195-5205.
102. Ward IM, Wu X, Chen J. Threonine 68 of Chk2 Is phosphorylated at sites of DNA strand breaks. J Biol Chem 2001; 276:47755-47758.
103. Buscemi G, Savio C, Zannini L et al. Chk2 activation dependence on NBS1 after DNA damage. Mol Cell Biol 2001; 21:5214-5222.
104. Girard PM, Riballo E, Begg AC et al. Nbs1 promotes ATM dependent phosphorylation events including those required for G1/S arrest. Oncogene 2002; 21:4191-4199.
105. Yarden RI, Pardo-Reoyo S, Sgagias M et al. BRCA1 regulates the G2/M checkpoint by activating Chk1 kinase upon DNA damage. Nat Genet 2002; 30:285-289.
106. Zachos G, Rainey MD, Gillespie DA. Chk1-deficient tumour cells are viable but exhibit multiple checkpoint and survival defects. EMBO J 2003; 22:713-723.
107. Gatei M, Sloper K, Sorensen C et al. Ataxia-telangiectasia-mutated (ATM) and NBS1-dependent phosphorylation of Chk1 on Ser-317 in response to ionizing radiation. J Biol Chem 2003; 278:14806-14811.
107a. Yamane K, Chen J, Kinsella TJ. Both DNA topoisomerase II-binding protein 1 and BRCA1 regulate the G2-M cell cycle checkpoint. Cancer Res 2003; 63:3049-3053.
108. Taalman RD, Jaspers NG, Scheres JM et al. Hypersensitivity to ionizing radiation, in vitro, in a new chromosomal breakage disorder, the Nijmegen Breakage Syndrome. Mutat Res 1983; 112:23-32.
109. Stewart GS, Maser RS, Stankovic T et al. The DNA double-strand break repair gene hMRE11 is mutated in individuals with an ataxia-telangiectasia-like disorder. Cell 1999; 99:577-587.
110. Kim YC, Koh JT, Shin BA et al. An antisense construct of full-length human RAD50 cDNA confers sensitivity to ionizing radiation and alkylating agents on human cell lines. Radiat Res 2002; 157:19-25.
111. Nelms BE, Maser RS, MacKay JF et al. In situ visualization of DNA double-strand break repair in human fibroblasts. Science 1998; 280:590-592.
112. Limoli CL, Ward JF. A new method for introducing double-strand breaks into cellular DNA. Radiat Res 1993; 134:160-169.
113. Maser RS, Monsen KJ, Nelms BE et al. hMre11 and hRad50 nuclear foci are induced during the normal cellular response to DNA double-strand breaks. Mol Cell Biol 1997; 17:6097-6104.
114. Mirzoeva OK, Petrini JH. DNA damage-dependent nuclear dynamics of the Mre11 complex. Mol Cell Biol 2001; 21:281-288.
115. Negorev D, Maul GG. Cellular proteins localized at and interacting within ND10/PML nuclear bodies/PODs suggest functions of a nuclear depot. Oncogene 2001; 20:7234-7242.
116. Lou Z, Chini CC, Minter-Dykhouse K et al. Mediator of DNA damage checkpoint protein 1 regulates BRCA1 localization and phosphorylation in DNA damage checkpoint control. J Biol Chem 2003; 278:13599-13602.
117. Paull TT, Cortez D, Bowers B et al. Direct DNA binding by Brca1. Proc Natl Acad Sci USA 2001; 98:6086-6091.
118. DeFazio LG, Stansel RM, Griffith JD et al. Synapsis of DNA ends by DNA-dependent protein kinase. EMBO J 2002; 21:3192-3200.
119. Lee SH, Kim CH. DNA-dependent protein kinase complex: A multifunctional protein in DNA repair and damage checkpoint. Mol Cells 2002; 13:159-166.
120. Chan DW, Chen BP, Prithivirajsingh S et al. Autophosphorylation of the DNA-dependent protein kinase catalytic subunit is required for rejoining of DNA double-strand breaks. Genes Dev 2002; 16:2333-2338.

121. Balajee AS, Geard CR. Chromatin-bound PCNA complex formation triggered by DNA damage occurs independent of the ATM gene product in human cells. Nucleic Acids Res 2001; 29:1341-1351.
122. Golub EI, Gupta RC, Haaf T et al. Interaction of human rad51 recombination protein with single-stranded DNA binding protein. RPA Nucleic Acids Res 1998; 26:5388-5393.
123. MacPhail SH, Olive PL. RPA foci are associated with cell death after irradiation. Radiat Res 2001; 155:672-679.
124. Yamane K, Kawabata M, Tsuruo T. A DNA-topoisomerase-II-binding protein with eight repeating regions similar to DNA-repair enzymes and to a cell-cycle regulator. Eur J Biochem 1997; 250:794-799.
125. Yamane K, Tsuruo T. Conserved BRCT regions of TopBP1 and of the tumor suppressor BRCA1 bind strand breaks and termini of DNA. Oncogene 1999; 18:5194-5203.
126. Makiniemi M, Hillukkala T, Tuusa J et al. BRCT domain-containing protein TopBP1 functions in DNA replication and damage response. J Biol Chem 2001; 276:30399-30406.
127. Yamane K, Wu X, Chen J. A DNA damage-regulated BRCT-containing protein, TopBP1, is required for cell survival. Mol Cell Biol 2002; 22:555-566.
128. DiBiase SJ, Zeng ZC, Chen R et al. DNA-dependent protein kinase stimulates an independently active, nonhomologous, end-joining apparatus. Cancer Res 2000; 60:1245-1253.
129. Stewart RD. Two-lesion kinetic model of double-strand break rejoining and cell killing. Radiat Res 2001; 156:365-378.
130. Thompson LH, Brookman KW, Dillehay LE et al. A CHO-cell strain having hypersensitivity to mutagens, a defect in strand-break repair, and an extraordinary baseline frequency of sister chromatid exchange. Mutat Res 1982; 95:427-440.
131. vanAnkeren SC, Murray D, Meyn RE. Induction and rejoining of ?-ray-induced DNA single- and double-strand breaks in Chinese hamster AA8 cells and in two radiosensitive clones. Radiat Res 1988; 116:511-525.
132. Nevaldine B, Longo JA, Hahn PJ. The scid defect results in much slower repair of DNA double-strand breaks but not high levels of residual breaks. Radiat Res 1997; 147:535-540.
133. Löbrich M, Rydberg B, Cooper PK. Repair of x-ray-induced DNA double-strand breaks in specific Not I restriction fragments in human fibroblasts: Joining of correct and incorrect ends. Proc Natl Acad Sci USA 1995; 92:12050-12054.
134. Rothkamm K, Kruger I, Thompson LH et al. Pathways of DNA double-strand break repair during the mammalian cell cycle. Mol Cell Biol 2003; in press.
135. Moynahan ME, Jasin M. Loss of heterozygosity induced by a chromosomal double-strand break. Proc Natl Acad Sci USA 1997; 94:8988-8993.
136. Moynahan ME, Pierce AJ, Jasin M. BRCA2 is required for homology-directed repair of chromosomal breaks. Mol. Cell 2001; 7:263-672.
137. Moynahan ME, Cui TY, Jasin M. Homology-directed DNA repair, mitomycin-C resistance, and chromosome stability is restored with correction of a Brca1 mutation. Cancer Res 2001; 61:4842-4850.
138. de Jager M, van Noort J, van Gent DC et al. Human Rad50/Mre11 is a flexible complex that can tether DNA ends. Mol Cell 2001; 8:1129-1135.
139. Hopfner KP, Craig L, Moncalian G et al. The Rad50 zinc-hook is a structure joining Mre11 complexes in DNA recombination and repair. Nature 2002; 418:562-566.
140. Wyman C, Kanaar R. Chromosome organization: Reaching out to embrace new models. Curr Biol 2002; 12:R446-R448.
141. Yang H, Jeffrey PD, Miller J et al. BRCA2 function in DNA binding and recombination from a BRCA2-DSS1-ssDNA structure. Science 2002; 297:1837-1848.
142. Kowalczykowski SC. Molecular mimicry connects BRCA2 to Rad51 and recombinational DNA repair. Nat Struct Biol 2002; 9:897-899.
143. Pellegrini L, Yu DS, Lo T et al. Insights into DNA recombination from the structure of a RAD51-BRCA2 complex. Nature 2002; 420:287-293.
144. Woodgate R. A plethora of lesion-replicating DNA polymerases. Genes Dev 1999; 13:2191-2195.
145. Patel PH, Loeb LA. Getting a grip on how DNA polymerases function. Nat Struct Biol 2001; 8:656-659.

146. Friedberg EC. Why do cells have multiple error-prone DNA polymerases? Environ Mol Mutagen 2001; 38:105-110.
147. Brown EJ, Baltimore D. ATR disruption leads to chromosomal fragmentation and early embryonic lethality. Genes Dev 2000; 14:397-402.
148. Xiao Y, Weaver DT. Conditional gene targeted deletion by Cre recombinase demonstrates the requirement for the double-strand break repair Mre11 protein in murine embryonic stem cells. Nucleic Acids Res 1997; 25:2985-2991.
149. Sonoda E, Sasaki M, Buerstedde JM et al. Rad51 deficient vertebrate cells accumulate chromosomal breaks prior to cell death. EMBO J 1998; 17:598-608.
150. Luo G, Yao MS, Bender CF et al. Disruption of mRad50 causes embryonic stem cell lethality, abnormal embryonic development, and sensitivity to ionizing radiation. Proc Natl Acad Sci USA 1999; 96:7376-7381.
151. Yamaguchi-Iwai Y, Sonoda E, Sasaki MS et al. Mre11 is essential for the maintenance of chromosomal DNA in vertebrate cells. EMBO J 1999; 18:6619-6629.
152. Zhu J, Petersen S, Tessarollo L et al. Targeted disruption of the Nijmegen breakage syndrome gene NBS1 leads to early embryonic lethality in mice. Curr Biol 2001; 11:105-109.
153. Tauchi H, Kobayashi J, Morishima K et al. Nbs1 is essential for DNA repair by homologous recombination in higher vertebrate cells. Nature 2002; 420:93-98.
154. Costanzo V, Robertson K, Bibikova M et al. Mre11 protein complex prevents double-strand break accumulation during chromosomal DNA replication. Mol Cell 2001; 8:137-147.
155. Lisby M, Rothstein R, Mortensen UH. Rad52 forms DNA repair and recombination centers during S phase. Proc Natl Acad Sci USA 2001; 98:8276-8282.
156. Bentley NJ, Holtzman DA, Flaggs G et al. The Schizosaccharomyces Pombe rad3 checkpoint gene. EMBO J 1996; 15:6641-6651.
157. Cimprich KA, Shin TB, Keith CT et al. cDNA cloning and gene mapping of a candidate human cell cycle checkpoint protein. Proc Natl Acad Sci USA 1996; 93:2850-2855.
158. von Hippel PH. The recombination-replication interface. Trends Biochem Sci 2000; 25:155.
159. Cox MM. Historical overview: Searching for replication help in all of the rec places. Proc Natl Acad Sci USA 2001; 98:8173-8180.
160. Nghiem P, Park PK, Kim Y et al. ATR inhibition selectively sensitizes G1 checkpoint-deficient cells to lethal premature chromatin condensation. Proc Natl Acad Sci USA 2001; 98:9092-9097.
161. Pierce AJ, Hu P, Han M et al. Ku DNA end-binding protein modulates homologous repair of double-strand breaks in mammalian cells. Genes Dev 2001; 15:3237-3242.
162. Cromie GA, Connelly JC, Leach DR. Recombination at double-strand breaks and DNA ends: Conserved mechanisms from phage to humans. Mol Cell 2001; 8:1163-1174.
163. Brown KD, Rathi A, Kamath R et al. The mismatch repair system is required for S-phase checkpoint activation. Nat Genet 2002; 33:80-84.
164. Painter RB, Young BR. Radiosensitivity in ataxia-telangiectasia: A new explanation. Proc Natl Acad Sci USA 1980; 77:7315-7317.
165. Costanzo V, Robertson K, Ying CY et al. Reconstitution of an ATM-dependent checkpoint that inhibits chromosomal DNA replication following DNA damage. Mol Cell 2000; 6:649-659.
166. Yazdi PT, Wang Y, Zhao S et al. SMC1 is a downstream effector in the ATM/NBS1 branch of the human S-phase checkpoint. Genes Dev 2002; 16:571-582.
167. Taniguchi T, Garcia-Higuera I, Xu B et al. Convergence of the Fanconi anemia and ataxia telangiectasia signaling pathways. Cell 2002; 109:459-472.
168. Falck J, Petrini JH, Williams BR et al. The DNA damage-dependent intra-S phase checkpoint is regulated by parallel pathways. Nat Genet 2002; 30:290-294.
169. Brown EJ, Baltimore D. Essential and dispensable roles of ATR in cell cycle arrest and genome maintenance. Genes Dev 2003; 17:615-628.
170. Walter J, Newport J. Initiation of eukaryotic DNA replication: Origin unwinding and sequential chromatin association of Cdc45, RPA and DNA polymerase alpha. Mol Cell 2000; 5:617-627.
171. Pages V, Fuchs RP. Uncoupling of leading- and lagging-strand DNA replication during lesion bypass in vivo. Science 2003; 300:1300-1303.
172. You Z, Kong L, Newport J. The role of single-stranded DNA and polymerase alpha in establishing the ATR, Hus1 DNA replication checkpoint. J Biol Chem 2002; 277:27088-27093.

173. Cordeiro-Stone M, Zaritskaya LS, Price LK et al. Replication fork bypass of a pyrimidine dimer blocking leading strand DNA synthesis. J Biol Chem 1997; 272:13945-13954.
174. Cordeiro-Stone M, Makhov AM, Zaritskaya LS et al. Analysis of DNA replication forks encountering a pyrimidine dimer in the template to the leading strand. J Mol Biol 1999; 289:1207-1218.
175. Wood RD. DNA repair. Variants on a theme. Nature 1999; 399:639-640.
176. Kiefer J, Feige M. The significance of DNA double-strand breaks in the UV inactivation of yeast cells. Mutat Res 1993; 299:219-224.
177. Limoli CL, Ward JF. Response of bromodeoxyuridine-substituted Chinese hamster cells to UVA light exposure in the presence of Hoechst dye #33258: Survival and DNA repair studies. Radiat Res 1994; 138:312-319.
178. Limoli CL, Giedzinski E, Morgan WF et al. Inaugural Article: Polymerase eta deficiency in the xeroderma pigmentosum variant uncovers an overlap between the S phase checkpoint and double-strand break repair. Proc Natl Acad Sci USA 2000; 97:7939-7946.
179. Limoli CL, Giedzinski E, Bonner WM et al. UV-induced replication arrest in the xeroderma pigmentosum variant leads to DNA double-strand breaks, gamma-H2AX formation, and Mre11 relocalization. Proc Natl Acad Sci USA 2002; 99:233-238.
180. Paull TT, Gellert M. The 3' to 5' exonuclease activity of Mre11 facilitates repair of DNA double-strand breaks. Mol Cell 1998; 1:969-979.
181. Paull TT, Gellert M. Nbs1 potentiates ATP-driven DNA unwinding and endonuclease cleavage by the Mre11/Rad50 complex. Genes Dev 1999; 13:1276-1288.
182. Lobachev KS, Gordenin DA, Resnick MA. The Mre11 complex is required for repair of hairpin-capped double-strand breaks and prevention of chromosome rearrangements. Cell 2002; 108:183-193.
183. Ma Y, Pannicke U, Schwarz K et al. Hairpin opening and overhang processing by an Artemis/DNA-dependent protein kinase complex in nonhomologous end joining and V(D)J recombination. Cell 2002; 108:781-794.
184. Postow L, Ullsperger C, Keller RW et al. Positive torsional strain causes the formation of a four-way junction at replication forks. J Biol Chem 2001; 276:2790-2796.
185. Sogo JM, Lopes M, Foiani M. Fork reversal and ssDNA accumulation at stalled replication forks owing to checkpoint defects. Science 2002; 297:599-602.
186. Tatsumi K, Strauss B. Production of DNA bifilarly substituted with bromodeoxyuridine in the first round of synthesis: Branch migration during isolation of cellular DNA. Nucleic Acids Res 1978; 5:331-347.
187. Tatsumi K, Strauss BS. Accumulation of DNA growing points in caffeine-treated human lymphoblastoid cells. J Mol Biol 1979; 135:435-449.
188. Mohaghegh P, Karow JK, Brosh Jr RM et al. The Bloom's and Werner's syndrome proteins are DNA structurespecific helicases. Nucleic Acids Res 2001; 29:2843-2849.
189. Mohaghegh P, Hickson ID. DNA helicase deficiencies associated with cancer predisposition and premature ageing disorders. Hum Mol Genet 2001; 10:741-746.
190. Chakraverty RK, Hickson ID. Defending genome integrity during DNA replication: A proposed role for RecQ family helicases. Bioessays 1999; 21:286-294.
191. Chen XB, Melchionna R, Denis CM et al. Human Mus81-associated endonuclease cleaves Holliday junctions in vitro. Mol Cell 2001; 8:1117-1127.
192. Haber JE, Heyer WD. The fuss about Mus81. Cell 2001; 107:551-554.
193. Michel B, Ehrlich SD, Uzest M. DNA double-strand breaks caused by replication arrest. EMBO J 1997; 16:430-408.
194. Bidnenko V, Ehrlich SD, Michel B. Replication fork collapse at replication terminator sequences. EMBO J 2002; 21:3898-3907.
195. Cha RS, Kleckner N. ATR homolog Mec1 promotes fork progression, thus averting breaks in replication slow zones. Science 2002; 297:602-606.
196. Arnaudeau C, Tenorio Miranda E, Jenssen D et al. Inhibition of DNA synthesis is a potent mechanism by which cytostatic drugs induce homologous recombination in mammalian cells. Mutat Res 2000; 461:221-228.

197. Lundin C, Erixon K, Arnaudeau C et al. Different roles for nonhomologous end joining and homologous recombination following replication arrest in mammalian cells. Mol Cell Biol 2002; 22:5869-5878.
198. Casper AM, Nghiem P, Arlt MF et al. ATR regulates fragile site stability. Cell 2002; 111:779-789.
199. Arnaudeau C, Lundin C, Helleday T. DNA double-strand breaks associated with replication forks are predominantly repaired by homologous recombination involving an exchange mechanism in mammalian cells. J Mol Biol 2001; 307:1235-1245.
200. Saintigny Y, Delacote F, Vares G et al. Characterization of homologous recombination induced by replication inhibition in mammalian cells. EMBO J 2001; 20:3861-3870.
201. Delacote F, Han M, Stamato TD et al. An xrcc4 defect or Wortmannin stimulates homologous recombination specifically induced by double-strand breaks in mammalian cells. Nucleic Acids Res 2002; 30:3454-3463.
202. Henry-Mowatt J, Jackson D, Masson JY et al. XRCC3 and Rad51 modulate replication fork progression on damaged vertebrate chromosomes. Mol Cell 2003; 11:1109-1117.
203. Liu N, Lamerdin JE, Tebbs RS et al. XRCC2 and XRCC3, new human Rad51-family members, promote chromosome stability and protect against DNA crosslinks and other damages. Mol Cell 1998; 1:783-793.
204. Saintigny Y, Lopez BS. Homologous recombination induced by replication inhibition, is stimulated by expression of mutant p53. Oncogene 2002; 21:488-492.
205. Gottifredi V, Shieh S, Taya Y et al. p53 accumulates but is functionally impaired when DNA synthesis is blocked. Proc Natl Acad Sci USA 2001; 98:1036-1041.
206. Cleaver J, Bartholomew J, Char D et al. Polymerase eta and p53 jointly regulate cell survival, apoptosis and Mre11 recombination during S phase checkpoint arrest after UV irradiation. DNA Repair 2001; 1:41-57.
207. Limoli CL, Laposa R, Cleaver JE. DNA replication arrest in XP variant cells after UV exposure is diverted into an Mre11-dependent recombination pathway by the kinase inhibitor wortmannin. Mutat Res 2002; 510:121-129.
208. Rouse J, Jackson SP. Interfaces between the detection, signaling, and repair of DNA damage. Science 2002; 297:547-551.
209. de Klein A, Muijtjens M, van Os R et al. Targeted disruption of the cell-cycle checkpoint gene ATR leads to early embryonic lethality in mice. Curr Biol 2000; 10:479-482.
210. Cortez D, Guntuku S, Qin J et al. ATR and ATRIP: Partners in checkpoint signaling. Science 2001; 294:1713-1716.
211. Cliby WA, Roberts CJ, Cimprich KA et al. Overexpression of a kinase-inactive ATR protein causes sensitivity to DNA-damaging agents and defects in cell cycle checkpoints. EMBO J 1998; 17:159-169.
212. Cliby WA, Lewis KA, Lilly KK et al. S phase and G2 arrests induced by topoisomerase I poisons are dependent on ATR kinase function. J Biol Chem 2002; 277:1599-1606.
213. Guo Z, Kumagai A, Wang SX et al. Requirement for Atr in phosphorylation of Chk1 and cell cycle regulation in response to DNA replication blocks and UV-damaged DNA in Xenopus egg extracts. Genes Dev 2000; 14:2745-2756.
214. Zhao H, Piwnica-Worms H. ATR-mediated checkpoint pathways regulate phosphorylation and activation of human Chk1. Mol Cel Biol 2001; 21:4129-4139.
215. Feijoo C, Hall-Jackson C, Wu R et al. Activation of mammalian Chk1 during DNA replication arrest: A role for Chk1 in the intra-S phase checkpoint monitoring replication origin firing. J Cell Biol 2001; 154:913-923.
216. Lupardus PJ, Byun T, Yee MC et al. A requirement for replication in activation of the ATR-dependent DNA damage checkpoint. Genes Dev 2002; 16:2327-2332.
217. Stokes MP, Van Hatten R, Lindsay HD et al. DNA replication is required for the checkpoint response to damaged DNA in Xenopus egg extracts. J Cell Biol 2002; 158:863-872.
218. Tibbetts RS, Brumbaugh KM, Williams JM et al. A role for ATR in the DNA damage-induced phosphorylation of p53. Genes Dev 1999; 13:152-157.
219. Nghiem P, Park PK, Kim Ys YS et al. ATR is not required for p53 activation but synergizes with p53 in the replication checkpoint. J Biol Chem 2002; 277:4428-4434.

220. Hammond EM, Denko NC, Dorie MJ et al. Hypoxia links ATR and p53 through replication arrest. Mol Cell Biol 2002; 22:1834-1843.
221. Tibbetts RS, Cortez D, Brumbaugh KM et al. Functional interactions between BRCA1 and the checkpoint kinase ATR during genotoxic stress. Genes Dev 2000; 14:2989-3002.
222. Gatei M, Zhou BB, Hobson K et al. Ataxia telangiectasia mutated (ATM) kinase and ATM and Rad3 related kinase mediate phosphorylation of Brca1 at distinct and overlapping sites. In vivo assessment using phospho-specific antibodies. J Biol Chem 2001; 276:17276-17280.
223. Hekmat-Nejad M, You Z, Yee M et al. Xenopus ATR is a replication-dependent chromatin-binding protein required for the DNA replication checkpoint. Curr Biol 2000; 10:1565-1573.
224. Kobayashi T, Tada S, Tsuyama T et al. Focus-formation of replication protein A, activation of checkpoint system and DNA repair synthesis induced by DNA double-strand breaks in Xenopus egg extract. J Cell Sci 2002; 115:3159-3169.
225. Michael WM, Ott R, Fanning E et al. Activation of the DNA replication checkpoint through RNA synthesis by primase. Science 2000; 289:2133-2137.
226. Takai H, Tominaga K, Motoyama N et al. Aberrant cell cycle checkpoint function and early embryonic death in Chk1(-/-) mice. Genes Dev 2000; 14:1439-1447.
227. Unsal-Kacmaz K, Makhov AM, Griffith JD et al. Preferential binding of ATR protein to UV-damaged DNA. Proc Natl Acad Sci USA 2002; 99:6673-6678.
228. Rouse J, Jackson SP. LCD1: An essential gene involved in checkpoint control and regulation of the MEC1 signalling pathway in Saccharomyces cerevisiae. EMBO J 2000; 19:5801-5812.
229. Edwards RJ, Bentley NJ, Carr AM. A Rad3-Rad26 complex responds to DNA damage independently of other checkpoint proteins. Nat Cell Biol 1999; 1:393-398.
230. Clerici M, Paciotti V, Baldo V et al. Hyperactivation of the yeast DNA damage checkpoint by TEL1 and DDC2 overexpression. EMBO J 2001; 20:6485-6498.
231. Rouse J, Jackson SP. Lcd1p recruits Mec1p to DNA lesions in vitro and in vivo. Mol Cell 2002; 9:857-869.
232. Green CM, Erdjument-Bromage H, Tempst P et al. A novel Rad24 checkpoint protein complex closely related to replication factor C. Curr Biol 2000; 10:39-42.
233. Rauen M, Burtelow MA, Dufault VM et al. The human checkpoint protein hRad17 interacts with the PCNA-like proteins hRad1, hHus1, and hRad9. J Biol Chem 2000; 275:29767-27971.
234. Lindsey-Boltz LA, Bermudez VP, Hurwitz J et al. Purification and characterization of human DNA damage checkpoint Rad complexes. Proc Natl Acad Sci USA 2001; 98:11236-11241.
235. Volkmer E, Karnitz LM. Human homologs of Schizosaccharomyces pombe Rad1, Hus1, and Rad9 form a DNA damage-responsive protein complex. J Biol Chem 1999; 274:567-570.
236. Hang H, Lieberman HB. Physical interactions among human checkpoint control proteins HUS1p, RAD1p, and RAD9p, and implications for the regulation of cell cycle progression. Genomics 2000; 65:24-33.
237. Burtelow MA, Roos-Mattjus PM, Rauen M et al. Reconstitution and molecular analysis of the hRad9-hHus1-hRad1 (9-1-1) DNA damage responsive checkpoint complex. J Biol Chem 2001; 276:25903-25909.
238. Shimomura T, Ando S, Matsumoto K et al. Functional and physical interaction between Rad24 and Rfc5 in the yeast checkpoint pathways. Mol Cell Biol 1998; 18:5485-5491.
239. Shimada M, Okuzaki D, Tanaka S et al. Replication factor C3 of Schizosaccharomyces pombe, a small subunit of replication factor C complex, plays a role in both replication and damage checkpoints. Mol Biol Cell 1999; 10:3991-4003.
240. Thelen MP, Venclovas C, Fidelis K. A sliding clamp model for the Rad1 family of cell cycle checkpoint proteins [letter]. Cell 1999; 96:769-770.
241. Venclovas C, Thelen MP. Structurebased predictions of rad1, rad9, hus1 and rad17 participation in sliding clamp and clamp-loading complexes. Nucleic Acids Res 2000; 28:2481-2493.
242. Caspari T, Dahlen M, Kanter-Smoler G et al. Characterization of Schizosaccharomyces pombe Hus1: A PCNA-related protein that associates with Rad1 and Rad9. Mol Cell Biol 2000; 20:1254-1262.
243. Griffith JD, Lindsey-Boltz LA, Sancar A. Structures of the human rad17-replication factor C and checkpoint rad 9-1-1 complexes visualized by glycerol spray/low voltage microscopy. J Biol Chem 2002; 277:15233-15236.

244. Roos-Mattjus P, Vroman BT, Burtelow MA et al. Genotoxin-induced Rad9-Hus1-Rad1 (9-1-1) chromatin association is an early checkpoint signaling event. J Biol Chem 2002; 277:43809-43812.
245. Wang X, Zou L, Zheng H et al. Genomic instability and endoreduplication triggered by RAD17 deletion. Genes Dev 2003.
246. Zou L, Cortez D, Elledge SJ. Regulation of ATR substrate selection by Rad17-dependent loading of Rad9 complexes onto chromatin. Genes Dev 2002; 16:198-208.
247. Kai M, Wang TS. Checkpoint activation regulates mutagenic translesion synthesis. Genes Dev 2003; 17:64-76.
248. Weiss RS, Matsuoka S, Elledge SJ et al. Hus1 acts upstream of Chk1 in a mammalian DNA damage response pathway. Curr Biol 2002; 12:73-77.
249. Weiss RS, Enoch T, Leder P. Inactivation of mouse Hus1 results in genomic instability and impaired responses to genotoxic stress. Genes Dev 2000; 14:1886-1898.
250. Hopkins KM, Auerbach W, Wang XY et al. Deletion of mouse Rad9 causes abnormal cellular responses to DNA damage, genomic instability and embryonic lethality. 2003; submitted.
251. Ward IM, Chen J. Histone H2AX is phosphorylated in an ATR-dependent manner in response to replicational stress. J Biol Chem 2001; 276:47759-47762.
252. Ishii Y, Bender MA. Effects of inhibitors of DNA synthesis on spontaneous and ultraviolet light-induced sister-chromatid exchanges in Chinese hamster cells. Mutat Res 1980; 79:19-32.
253. Sonoda E, Sasaki MS, Morrison C et al. Sister chromatid exchanges are mediated by homologous recombination in vertebrate cells. Mol Cell Biol 1999; 19:5166-5169.
254. Mirzoeva OK, Petrini JH. DNA replication-dependent nuclear dynamics of the Mre11 complex. Mol Cancer Res 2003; 1:207-218.
255. Fenech M, Carr AM, Murray J et al. Cloning and characterization of the rad4 gene of Schizosaccharomyces pombe; a gene showing short regions of sequence similarity to the human XRCC1 gene. Nucleic Acids Res 1991; 19:6737-7641.
256. Araki H, Leem SH, Phongdara A et al. Dpb11, which interacts with DNA polymerase II(epsilon) in Saccharomyces cerevisiae, has a dual role in S-phase progression and at a cell cycle checkpoint. Proc Natl Acad Sci USA 1995; 92:11791-11795.
257. Yamamoto RR, Axton JM, Yamamoto Y et al. The Drosophila mus101 gene, which links DNA repair, replication and condensation of heterochromatin in mitosis, encodes a protein with seven BRCA1 C-terminus domains. Genetics 2000; 156:711-721.
258. Wang H, Elledge SJ. DRC1, DNA replication and checkpoint protein 1, functions with DPB11 to control DNA replication and the S-phase checkpoint in Saccharomyces cerevisiae. Proc Natl Acad Sci USA 1999; 96:3824-3829.
259. McFarlane RJ, Carr AM, Price C. Characterisation of the Schizosaccharomyces pombe rad4/cut5 mutant phenotypes: Dissection of DNA replication and G2 checkpoint control function. Mol Gen Genet 1997; 255:332-340.
260. Saka Y, Esashi F, Matsusaka T et al. Damage and replication checkpoint control in fission yeast is ensured by interactions of Crb2, a protein with BRCT motif, with Cut5 and Chk1. Genes Dev 1997; 11:3387-3400.
261. Saka Y, Fantes P, Sutani T et al. Fission yeast cut5 links nuclear chromatin and M phase regulator in the replication checkpoint control. EMBO J 1994; 13:5319-5329.
262. van Brabant AJ, Stan R, Ellis NA. DNA helicases, genomic instability, and human genetic disease. Annu Rev Genomics Hum Genet 2000; 1:409-459.
263. Nakayama H. RecQ family helicases: Roles as tumor suppressor proteins. Oncogene 2002; 21:9008-9021.
264. Gangloff S, McDonald JP, Bendixen C et al. The yeast type I topoisomerase Top3 interacts with Sgs1, a DNA helicase homolog: A potential eukaryotic reverse gyrase. Mol. Cell Biol 1994; 14:8391-8398.
265. Watt PM, Louis EJ, Borts RH et al. Sgs1: A eukaryotic homolog of E. coli RecQ that interacts with topoisomerase II in vivo and is required for faithful chromosome segregation. Cell 1995; 81:253-260.
266. Stewart E, Chapman CR, Al-Khodairy F et al. rqh1+, a fission yeast gene related to the Bloom's and Werner's syndrome genes, is required for reversible S phase arrest. EMBO J 1997; 16:2682-2692.

267. Murray JM, Lindsay HD, Munday CA et al. Role of Schizosaccharomyces pombe RecQ homolog, recombination, and checkpoint genes in UV damage tolerance. Mol Cell Biol 1997; 17:6868-6875.
268. Ellis NA, Groden J, Ye TZ et al. The Bloom's syndrome gene product is homologous to RecQ helicases. Cell 1995; 83:655-666.
269. Yu CE, Oshima J, Fu YH et al. Positional cloning of the Werner's syndrome gene. Science 1996; 272:258-262.
270. Seki M, Miyazawa H, Tada S et al. Molecular cloning of cDNA encoding human DNA helicase Q1 which has homology to Escherichia coli Rec Q helicase and localization of the gene at chromosome 12p12. Nucleic Acids Res 1994; 22:4566-4573.
271. Puranam KL, Blackshear PJ. Cloning and characterization of RECQL, a potential human homologue of the Escherichia coli DNA helicase RecQ. J Biol Chem 1994; 269:29838-29845.
272. Kitao S, Ohsugi I, Ichikawa K et al. Cloning of two new human helicase genes of the RecQ family: Biological significance of multiple species in higher eukaryotes. Genomics 1998; 54:443-452.
273. Sekelsky JJ, Brodsky MH, Rubin GM et al. Drosophila and human RecQ5 exist in different isoforms generated by alternative splicing. Nucleic Acids Res 1999; 27:3762-3769.
274. Shimamoto A, Nishikawa K, Kitao S et al. Human RecQ5beta, a large isomer of RecQ5 DNA helicase, localizes in the nucleoplasm and interacts with topoisomerases 3alpha and 3beta. Nucleic Acids Res 2000; 28:1647-1655.
275. Kitao S, Lindor NM, Shiratori M et al. Rothmund-Thomson syndrome responsible gene, RECQL4: Genomic structure and products. Genomics 1999; 61:268-276.
276. Kitao S, Shimamoto A, Goto M et al. Mutations in RECQL4 cause a subset of cases of Rothmund-Thomson syndrome. Nat Genet 1999; 22:82-84.
277. Balraj P, Concannon P, Jamal R et al. An unusual mutation in RECQ4 gene leading to Rothmund-Thomson syndrome. Mutat Res 2002; 508:99-105.
278. Sengupta S, Linke SP, Pedeux R et al. BLM helicase-dependent transport of p53 to sites of stalled DNA replication forks modulates homologous recombination. EMBO J 2003; 22:1210-1222.
279. Lindor NM, Furuichi Y, Kitao S et al. Rothmund-Thompson syndrome due to RECQ4 helicase mutations: Report and clinical and molecular comparisons with Bloom syndrome and Werner syndrome. Am J Med Genet 2000; 90:223-228.
280. Shen J, Loeb LA. Unwinding the molecular basis of the Werner syndrome. Mech Ageing Dev 2001; 122:921-944.
281. Brosh Jr RM, Bohr VA. Roles of the Werner syndrome protein in pathways required for maintenance of genome stability. Exp Gerontol 2002; 37:491-506.
282. Mohaghegh P, Hickson I. Premature aging in RecQ helicase-deficient human syndromes. Int J Biochem Cell Biol 2002; 34:1496-1501.
283. Gianneli F, Benson PF, Pawsey SA et al. Ultraviolet light sensitivity and delayed DNA-chain maturation in Bloom's syndrome fibroblasts. Nature 1977; 265:466-469.
284. Hanaoka F, Yamada M, Takeuchi F et al. Autoradiographic studies of DNA replication in the Werner syndrome. In: Salk D, Fujiwara Y, Martin GM, eds. Werner's syndrome and human aging. New York: Plenum Press, 1985:439-457.
285. Lonn U, Lonn S, Nylen U et al. An abnormal profile of DNA replication intermediates in Bloom's syndrome. Cancer Res 1990; 50:3141-3145.
286. Chaganti RSK, Schonberg S, German J. A manyfold increase in sister chromatid exchanges in Bloom's syndrome lymphocytes. Proc Natl Acad Sci USA 1974; 71:4508-4512.
287. Saintigny Y, Makienko K, Swanson C et al. Homologous recombination resolution defect in Werner syndrome. Mol Cell Biol 2002; 22:6971-6978.
288. Fukuchi K, Martin GM, Monnat Jr RJ. Mutator phenotype of Werner syndrome is characterized by extensive deletions. Proc Natl Acad Sci USA 1989; 86:5893-5897. [published erratum appears in Proc. Natl. Acad. Sci. USA 1989, 86:7994]
289. Lebel M. Increased frequency of DNA deletions in pink-eyed unstable mice carrying a mutation in the Werner syndrome gene homologue. Carcinogenesis 2002; 23:213-216.
290. Yamagata K, Kato J, Shimamoto A et al. Bloom's and Werner's syndrome genes suppress hyperrecombination in yeast sgs1 mutant: Implication for genomic instability in human disease. Proc Natl Acad Sci USA 1998; 95:8733-8738.

291. Imamura O, Fujita K, Shimamoto A et al. Bloom helicase is involved in DNA surveillance in early S phase in vertebrate cells. Oncogene 2001; 20:1143-1151.
292. Franchitto A, Pichierri P. Bloom's syndrome protein is required for correct relocalization of RAD50/MRE11/NBS1 complex after replication fork arrest. J Cell Biol 2002; 157:19-30.
293. Franchitto A, Pichierri P. Protecting genomic integrity during DNA replication: Correlation between Werner's and Bloom's syndrome gene products and the MRE11 complex. Hum Mol Genet 2002; 11:2447-2453.
294. Ababou M, Dumaire V, Lecluse Y et al. Bloom's syndrome protein response to ultraviolet-C radiation and hydroxyurea-mediated DNA synthesis inhibition. Oncogene 2002; 21:2079-2088.
295. Okada M, Goto M, Furuichi Y et al. Differential effects of cytotoxic drugs on mortal and immortalized B-lymphoblastoid cell lines from normal and Werner's syndrome patients. Biol Pharm Bull 1998; 21:235-239.
296. Lebel M, Leder P. A deletion within the murine Werner syndrome helicase induces sensitivity to inhibitors of topoisomerase and loss of cellular proliferative capacity. Proc Natl Acad Sci USA 1998; 95:13097-13102.
297. Poot M, Gollahon KA, Rabinovitch PS. Werner syndrome lymphoblastoid cells are sensitive to camptothecin-induced apoptosis in S-phase. Hum Genet 1999; 104:10-14.
298. Pichierri P, Franchitto A, Mosesso P et al. Werner's syndrome cell lines are hypersensitive to camptothecin-induced chromosomal damage. Mutat Res 2000; 456:45-57.
299. Pichierri P, Franchitto A, Mosesso P et al. Werner's syndrome protein is required for correct recovery after replication arrest and DNA damage induced in S-phase of cell cycle. Mol Biol Cell 2001; 12:2412-2421.
300. Karow JK, Constantinou A, Li JL et al. The Bloom's syndrome gene product promotes branch migration of holliday junctions. Proc Natl Acad Sci USA 2000; 97:6504-6508.
301. Constantinou A, Tarsounas M, Karow JK et al. Werner's syndrome protein (WRN) migrates holliday junctions and colocalizes with RPA upon replication arrest. EMBO Rep 2000; 1:80-84.
302. Yang Q, Zhang R, Wang XW et al. The processing of Holliday junctions by BLM and WRN helicases is regulated by p53. J Biol Chem 2002; 277:31980-31987.
303. Wang W, Seki M, Narita Y et al. Possible association of BLM in decreasing DNA double strand breaks during DNA replication. EMBO J 2000; 19:3428-3435.
304. Ng SW, Liu Y, Hasselblatt KT et al. A new human topoisomerase III that interacts with SGS1 protein. Nucleic Acids Res 1999; 27:993-1000.
305. Wu L, Davies SL, North PS et al. The Bloom's syndrome gene product interacts with topoisomerase III. J Biol Chem 2000; 275:9636-9644.
306. Johnson FB, Lombard DB, Neff NF et al. Association of the Bloom syndrome protein with topoisomerase IIIalpha in somatic and meiotic cells. Cancer Res 2000; 60:1162-1167.
307. Hyde H, Davies AA, Benson FE et al. Resolution of recombination intermediates by a mammalian activity functionally analogous to Escherichia coli RuvC resolvase. J Biol Chem 1994; 269:5202-5209.
308. Constantinou A, Davies AA, West SC. Branch migration and Holliday junction resolution catalyzed by activities from mammalian cells. Cell 2001; 104:259-268.
309. Kaliraman V, Mullen JR, Fricke WM et al. Functional overlap between Sgs1-Top3 and the Mms4-Mus81 endonuclease. Genes Dev 2001; 15:2730-2740.
310. Boddy MN, Lopez-Girona A, Shanahan P et al. Damage tolerance protein Mus81 associates with the FHA1 domain of checkpoint kinase Cds1. Mol Cell Biol 2000; 20:8758-8566.
311. Doe CL, Ahn JS, Dixon J et al. Mus81-Eme1 and Rqh1 involvement in processing stalled and collapsed replication forks. J Biol Chem 2002; 277:32753-32759.
312. Beamish H, Kedar P, Kaneko H et al. Functional link between BLM defective in Bloom's syndrome and the ataxia-telangiectasia mutated protein, ATM. J Biol Chem 2002; 277:30515-30523.
313. Cortez D, Wang Y, Qin J et al. Requirement of ATM-dependent phosphorylation of Brca1 in the DNA damage response to double-strand breaks. Science 1999; 286:1162-1166.
314. Gatei M, Scott SP, Filippovitch I et al. Role for ATM in DNA damage-induced phosphorylation of BRCA1. Cancer Res 2000; 60:3299-3304.
315. Xu B, Kim ST, Kastan MB. Involvement of Brca1 in S-phase and G(2)-phase checkpoints after ionizing irradiation. Mol Cell Biol 2001; 21:3445-3450.

316. Xu B, O'Donnell AH, Kim ST et al. Phosphorylation of serine 1387 in Brca1 is specifically required for the Atm-mediated S-phase checkpoint after ionizing irradiation. Cancer Res 2002; 62:4588-4591.
317. Okada S, Ouchi T. Cell cycle differences in DNA-damage-induced BRCA1 phosphorylation affect its subcellular localization. J Biol Chem 2003.
318. Khanna KK, Keating KE, Kozlov S et al. ATM associates with and phosphorylates p53: Mapping the region of interaction. Nat Genet 1998; 20:398-400.
319. Chehab NH, Malikzay A, Appel M et al. Chk2/hCds1 functions as a DNA damage checkpoint in G(1) by stabilizing p53. Genes Dev 2000; 14:278-288.
320. Matsuoka S, Rotman G, Ogawa A et al. Ataxia telangiectasia-mutated phosphorylates chk2 in vivo and in vitro. Proc Natl Acad Sci USA 2000; 97:10389-01394.
321. Melchionna R, Chen XB, Blasina A et al. Threonine 68 is required for radiation-induced phosphorylation and activation of Cds1. Nat Cell Biol 2000; 2:762-765.
322. Ahn JY, Schwarz JK, Piwnica-Worms H et al. Threonine 68 phosphorylation by ataxia telangiectasia mutated is required for efficient activation of Chk2 in response to ionizing radiation. Cancer Res 2000; 60:5934-5936.
323. Xu X, Tsvetkov LM, Stern DF. Chk2 activation and phosphorylation-dependent oligomerization. Mol Cell Biol 2002; 22:4419-4432.
324. Ahn J, Prives C. Checkpoint Kinase 2 (Chk2) Monomers or Dimers Phosphorylate Cdc25C after DNA Damage Regardless of Threonine 68 Phosphorylation. J Biol Chem 2002; 277:48418-48426.
325. Kim ST, Xu B, Kastan MB. Involvement of the cohesin protein, Smc1, in Atm-dependent and independent responses to DNA damage. Genes Dev 2002; 16:560-570.
326. Chen MJ, Lin YT, Lieberman HB et al. Atm-dependent phosphorylation of human rad9 is required for ionizing radiation-induced checkpoint activation. J Biol Chem 2001; 276:16580-16586.
327. Gately DP, Hittle JC, Chan GK et al. Characterization of ATM expression, localization, and associated DNA-dependent protein kinase activity. Mol Biol Cell 1998; 9:2361-2374.
328. Chan DW, Son SC, Block W et al. Purification and characterization of ATM from human placenta. A manganese-dependent, wortmannin-sensitive serine/threonine protein kinase. J Biol Chem 2000; 275:7803-7810.
329. Wang H, Guan J, Wang H et al. Replication protein A2 phosphorylation after DNA damage by the coordinated action of ataxia telangiectasia-mutated and DNA-dependent protein kinase. Cancer Res 2001; 61:8554-8563.
330. Sapkota GP, Deak M, Kieloch A et al. Ionizing radiation induces ataxia telangiectasia mutated kinase (ATM)-mediated phosphorylation of LKB1/STK11 at Thr-366. Biochem J 2002; 368:507-516.
331. Stewart GS, Last JI, Stankovic T et al. Residual ataxia telangiectasia mutated protein function in cells from ataxia telangiectasia patients, with 5762ins137 and 7271T—>G mutations, showing a less severe phenotype. J Biol Che 2001; 276:30133-30141.
332. Lee JS, Collins KM, Brown AL et al. hCds1-mediated phosphorylation of BRCA1 regulates the DNA damage response. Nature 2000; 404:201-204.
333. Wu L, Davies SL, Levitt NC et al. Potential role for the BLM helicase in recombinational repair via a conserved interaction with RAD51. J Biol Chem 2001; 276:19375-19381.
334. Davies AA, Masson JY, McIlwraith MJ et al. Role of BRCA2 in control of the RAD51 recombination and DNA repair protein. Mol Cell 2001; 7:273-282.
335. Wu LC, Wang ZW, Tsan JT et al. Identification of a RING protein that can interact in vivo with the BRCA1 gene product. Nat Genet 1996; 14:430-440.
336. Chen J, Silver DP, Walpita D et al. Stable interaction between the products of the BRCA1 and BRCA2 tumor suppressor genes in mitotic and meiotic cells. Mol Cell 1998; 2:317-328.
337. Chai YL, Cui J, Shao N et al. The second BRCT domain of BRCA1 proteins interacts with p53 and stimulates transcription from the p21WAF1/CIP1 promoter. Oncogene 1999; 18:263-268.
338. Zhong Q, Chen CF, Li S et al. Association of BRCA1 with the hRad50-hMre11-p95 complex and the DNA damage response. Science 1999; 285:747-750.
339. Garcia-Higuera I, Taniguchi T, Ganesan S et al. Interaction of the Fanconi anemia proteins and BRCA1 in a common pathway. Mol Cell 2001; 7:249-262.

340. Fujimori A, Tachiiri S, Sonoda E et al. Rad52 partially substitutes for the Rad51 paralog XRCC3 in maintaining chromosomal integrity in vertebrate cells. EMBO J 2001; 20:5513-5520.
341. Lakin ND, Hann BC, Jackson SP. The ataxia-telangiectasia related protein ATR mediates DNA-dependent phosphorylation of p53. Oncogene 1999; 18:3989-3995.
342. Liu Q, Guntuku S, Cui XS et al. Chkl is an essential kinase that is regulated by ATR and required for the G(2)/M DNA damage checkpoint. Genes Dev 2000; 14:1448-1459.
343. Bao S, Tibbetts RS, Brumbaugh KM et al. ATR/ATM-mediated phosphorylation of human Rad17 is required for genotoxic stress responses. Nature 2001; 411:969-974.
344. Wang X, Wang L, Callister MD et al. Human Rad17 is phosphorylated upon DNA damage and also overexpressed in primary nonsmall cell lung cancer tissues. Cancer Res 2001; 61:7417-7421.
345. Post S, Weng YC, Cimprich K et al. Phosphorylation of serines 635 and 645 of human Rad17 is cell cycle regulated and is required for G1/S checkpoint activation in response to DNA damage. Proc Natl Acad Sci USA 2001; 98:13102-13107.
346. Karmakar P, Piotrowski J, Brosh Jr RM et al. Werner protein is a target of DNA-PK in vivo and in vitro, and its catalytic activities are regulated by phosphorylation. J Biol Chem 2002; 277:18291-18302.
347. Wang XW, Tseng A, Ellis NA et al. Functional interaction of p53 and BLM DNA helicase in apoptosis. J Biol Chem 2001; 276:32948-32955.
348. Garkavtsev IV, Kley N, Grigorian IA et al. The Bloom syndrome protein interacts and cooperates with p53 in regulation of transcription and cell growth control. Oncogene 2001; 20:8276-8280.
349. Von Kobbe C, Karmakar P, Dawut L et al. Colocalization, physical, and functional interaction between Werner and Bloom syndrome proteins. J Biol Chem 2002; 277:22035-22044.
350. Brosh J, Karmakar P, Sommers JA et al. p53 modulates the exonuclease activity of Werner syndrome protein. J Biol Chem 2001; 276:35093-35102.
351. Baynton K, Otterlei M, Bjoras M et al. WRN interacts physically and functionally with the recombination mediator protein Rad52. J Biol Chem 2003; in press.
352. Dahm K, Hubscher U. Colocalization of human Rad17 and PCNA in late S phase of the cell cycle upon replication block. Oncogene 2002; 21:7710-7719.
353. Sagata N. MOLECULAR BIOLOGY: Untangling checkpoints. Science 2002; 298:1905-1907.
354. Zhao H, Watkins JL, Piwnica-Worms H. Disruption of the checkpoint kinase 1/cell division cycle 25A pathway abrogates ionizing radiation-induced S and G2 checkpoints. Proc Natl Acad Sci USA 2002; 99:14795-14800.
355. Donzelli M, Squatrito M, Ganoth D et al. Dual mode of degradation of Cdc25A phosphatase. EMBO J 2002; 21:4875-4884.
356. Foray N, Marot D, Gabriel A et al. A subset of ATM- and ATR-dependent phosphorylation events requires the BRCA1 protein. EMBO J 2003; 11:2860-2871.

CHAPTER 7

The Mechanism of DNA Non-Homologous End-Joining:

Lessons Learned from Biophysical, Biochemical and Cellular Studies

P.A. Jeggo

Abstract

DNA non-homologous end-joining (NHEJ) is a major mechanism for the repair of DNA double strand breaks (DSBs). The initial discovery that the Ku protein plays a key role in NHEJ was made in 1994 and in the ensuing few years the DNA-dependent protein kinase catalytic subunit (DNA-PKcs), XRCC4 and DNA ligase IV were all shown to be essential components. Since this time, a diverse range of studies involving structural, biochemical and cellular approaches, have provided considerable insight into the roles of the individual proteins and the nature of the overall process. Here, the findings from these differing disciplines are reviewed and a model is presented for NHEJ and its role in V(D)J recombination.

Introduction

A DNA double strand break (DSB) is a profoundly dangerous lesion to a cell, which, if unrepaired, will result in loss of genomic material. A misrepaired DSB can also have serious consequences for a cell and, just as importantly, for an organism, by leading to rearrangements as well as loss of genetic material. Cells have invested in at least two pathways to repair DSBs, namely homologous recombination (HR) and non-homologous end-joining (NHEJ). This chapter will focus on NHEJ, a mechanism that rejoins DSBs using little or no homology.

In higher organisms, NHEJ represents the major mechanism for the repair of radiation induced DSBs, as demonstrated by the dramatic radiosensitivity of cell lines lacking components of the NHEJ machinery (for reviews see refs. 1,2) Such defective cells also display genomic instability suggesting that NHEJ also plays at least some role in the repair of endogenously arising DSBs.[3] A plethora of evidence has also demonstrated that NHEJ functions during the process of V(D)J recombination to effect site specific recombination at the DSBs that arise following the combined action of the recombination activating gene proteins, RAG1 and RAG2.[4,5] Mice and humans lacking or impaired in components of the NHEJ machinery thus display a severe combined immunodeficiency (SCID) phenotype (for a review see ref. 6). More recently, NHEJ has been implicated in class switch recombination, another rearrangement process that helps to enhance diversity during immune development.[7-9] Finally, at least

Eukaryotic DNA Damage Surveillance and Repair, edited by Keith W. Caldecott.

some of the proteins required for NHEJ are found at telomeres and play a role in telomere maintenance (see for example ref. 10).

Five core components (Ku70, Ku80, DNA-PKcs, XRCC4 and DNA ligase IV) of the NHEJ machinery have been identified in mammalian cells and some further additional proteins have been shown to impact upon the process. A role for Ku, a heterodimer with subunits of 70 and 80 kDa, in NHEJ was recognised in 1994 with the identification of one of the subunits, Ku80, as the gene defective in a rodent cell line that displayed radiosensitivity and a DSB rejoining defect.[11,12] Furthermore, since Ku had been previously shown to associate with a large protein, the DNA dependent protein kinase catalytic subunit (DNA-PKcs), to generate the DNA dependent protein kinase (DNA-PK) complex, DNA-PKcs was also implicated as a likely protein participating in the process.[13,14] The identification of DNA-PKcs as the gene defective in the SCID mouse, whose cells also showed radiosensitivity, consolidated this notion.[15] The identification of XRCC4 and DNA ligase IV as two further components of the same pathway rapidly followed.[16-18] The DSB rejoining pathway involving these proteins soon became known as DNA NHEJ. Since other pathways of end-joining that require little or no homology also clearly exist, the pathway under discussion here has also, more precisely, been called DNA-PK dependent NHEJ. Since this time, although no quantum advances in our understanding of the process have been made, considerable insight into the mechanism has been gained from a range of structural, biochemical and cellular experiments. The aim of this chapter is to provide a current over-view of the process of NHEJ focusing on recent advances gained from these divergent disciplines.

Nature of the DNA Ends to Be Rejoined

Since the end-processing events required prior to ligation will determine the additional proteins required for NHEJ, the nature of the DNA ends generated by DNA damaging agents and other metabolic events leading to DSB formation is important in considering the overall process of NHEJ. DNA ligases require 3'-hydroxyl (3' OH) and 5'-phosphate (5'P) termini for ligation. Such termini are normally generated by restriction endonucleases and may arise during some other forms of DNA repair processing. However, IR, in particular, as well as other forms of DNA damaging agents including endogenously arising reactive oxygen species, generate 3'-monophosphate or 3'-phosphoglycolate end groups. 5'OH and 5'-deoxyribose phosphate end groups also arise frequently. Significantly, XRCC1, a protein involved in single strand break (SSB) repair, interacts with polynucleotide kinase (PNK) and stimulates both its DNA kinase and DNA phosphatase activities at damaged termini, providing one mechanism to aid the rejoining of single strand breaks with termini that are not directly amenable to ligation.[19] More recently, PNK has also been implicated to aid the rejoining of non-ligatable ends during NHEJ.[20] Depending on the type of radiation, many DSBs have DNA ends associated with damaged bases, which may also require removal prior to rejoining.

Most importantly, the coding ends generated during V(D)J recombination require specific modification. During the process of V(D)J recombination, DSBs are generated by the combined action of two proteins encoded by the recombination activating genes, *RAG1* and *2*.[21] The initial step of the process is an endonucleolytic cleavage event between a coding sequence and its adjacent signal sequence. This is rapidly followed by nucleophilic attack of the 3'OH end formed on the phosphodiester bond in the opposite strand leading to the formation of a hairpin DNA end at the coding sequence and a blunt DNA end at the signal sequence.[5,22] The hairpin end requires resolution prior to formation of an effective V(D)J junction. Rejoining of the blunt ended signal junction is not, per se, essential for V(D)J recombination, although an unrejoined DSB may lead to cell death via apoptosis. A consideration of the factors required for these different rejoining events will be discussed in further detail below.

Structure and Biochemical Analysis of the Core NHEJ Proteins

Ku

Ku has two striking properties; firstly, it is one of the strongest double-stranded (ds) DNA end-binding proteins described to date with little, if any, ability to bind to circular DNA;[23] secondly, Ku has the ability to translocate along the DNA in an energy independent manner[23,24] (for a review see ref. 25). Insight into both of these properties was provided by the recent resolution of the crystal structure.[26] Ku70 and Ku80 show considerable homology and it has been suggested that they originate from a common ancestor. Ku80 has a C-terminal domain that does not show any sequence homology with Ku70 but is required for interaction with DNA-PKcs.[27-29] Yeast Ku80 has a shorter C-terminus compared to mammalian homologues and significantly, non-vertebrates do not have a DNA-PKcs homologue consistent with the notion that the C-terminus of Ku80 is required for interaction with DNA-PKcs. Crystallisation was carried out on a Ku70/80 heterodimer that lacked the Ku80 C-terminus and on a dimer bound to a short DNA oligonucleotide.[26] The structure consolidated the relatedness between the two Ku subunits since the fold of Ku70 and Ku80 are closely related and form a structure with a high degree of symmetry consistent with their divergence from a common ancestor. Furthermore, the complex fold of the two subunits is consistent with earlier evidence that mutations at multiple sites of Ku80 impair heterodimer formation.[30] More significantly, the structure resembles a basket shape with a larger base that serves to cradle the DNA, two pillars and a bridge. Together, these form a loop through which the DNA can pass. The binding site cradles two turns of DNA but encircles only the central 3-4 base pairs with no direct contact being made with the DNA bases. The structure explains the requirement of a free ds DNA end for Ku binding, the lack of any sequence specificity and suggests that the interaction of Ku with the DNA allows the exposure of a region of DNA at the end to facilitate binding of other molecules involved in the rejoining process. The structure, additionally, shows a preferred orientation for DNA binding and suggested that inward movement of Ku can occur but with an energetic barrier consistent with the inward translocation of Ku representing a rate-limiting step. Biophysical studies, including the use of dynamic light scattering, double-filter binding assays and fluorescence spectroscopy have also provided important information on the dynamics of the interaction between Ku and DNA (see for example ref. 31). Such studies have shown, additionally, that the binding of multiple Ku molecules to the DNA occurs in a non-co-operative manner but that the Ku molecule moves along the DNA without interfering with the binding of additional Ku molecules to the same DNA end.

Another feature reported for Ku is an ability to bridge two DNA molecules and thereby stabilize an intermolecular association of DNA ends.[32] The ability of Ku to bridge DNA ends is also suggested from atomic force microscopy studies showing that Ku-bound DNA ends co-associate to cause DNA loop formation.[33] It has also been shown that Ku can transfer efficiently between two DNA molecules.[34] The impact of these properties of Ku on NHEJ and the ability of Ku to interact with other components of the DNA NHEJ machinery will be discussed below.

DNA-PKcs

Once Ku is bound to DNA ends it can associate with DNA-PKcs to form the DNA-PK complex.[13,14,25] DNA-PKcs belongs to a sub-family of phosphoinositol 3-kinases called phosphoinsolitol 3-kinase like kinases (PIKKs). It is a serine-threonine protein kinase, and at 465 kDa is one of the largest family members. Its kinase activity is activated by binding to Ku and is, therefore, activated by DNA ends. The mutation in the SCID mouse results in loss of the extreme C-terminal region of DNA-PKcs.[35] The truncated protein is unable to function as a protein kinase demonstrating that the extreme C-terminus of the PIKK family, which is not found in PI 3-kinases with lipid kinase activity, is required for these proteins to function as protein kinases.[36] The DSB repair defect of the SCID mouse strongly suggested that the kinase function of DNA-PKcs is required for NHEJ. This has been consolidated more recently, with the demonstration that a kinase dead form of DNA-PKcs is unable to rescue either the radiosensitivity or DSB rejoining defect of a cell line defective in DNA-PKcs.[37] It has also been shown that DNA-PKcs undergoes autophosphorylation and that a single phosphorylation site (Thr2609) in DNA-PKcs is required for its function in NHEJ.[38] Furthermore, in separate studies phosphorylation of DNA-PK has been shown to lead to disruption of the DNA-PK/DNA complex.[39] Together these results suggest that, not only is the kinase function essential for NHEJ, but autophosphorylation of DNA-PKcs is also required.

DNA-PKcs forms a stable complex with DNA even in the absence of Ku and has been reported to be activated by DNA in the absense of Ku under certain conditions.[40,41] UV photo-cross linking experiments have shown that DNA-PKcs makes direct contact with a DNA terminus and occupies about 10 bp proximal to the DNA end.[14,42] More significantly, the recruitment of DNA-PKcs to a DNA molecule with Ku bound at the end induces inward translocation of Ku by approximately one helical turn, a finding that could represent the first step of the NHEJ reaction.[42] An important question is whether Ku alone or the DNA-PK complex functions to hold DNA ends together. Ku has been shown to transfer between DNA ends although this does not necessitate tethering of DNA ends.[34] However, biochemical studies complemented with electron microscopic analysis have provided evidence that either Ku alone or together with DNA-PKcs helps to tether DNA ends thereby mediating the formation of a synaptic complex containing two DNA molecules.[33,43] Significantly, whilst the kinase activity appears to be dispensible for synapse formation, it is activated following synapsis formation raising the possibility that DNA-PKcs may function to bring ends together and then undergo autophosphorylation with subsequent release from the DNA end.

DNA Ligase IV/XRCC4 (LX) Complex

DNA ligase IV forms an exceedingly tight complex with XRCC4 such that the complex can only be forced to dissociate in vitro at very high salt concentrations.[17,44] The N-terminal domain of ligase IV has a region that shares structural homology with other ATP-dependent DNA ligases.[45-47] This encompasses an adenylation domain and a smaller, OB domain, found in many nucleic acid binding proteins.[48] Within these regions, five highly conserved motifs (I-V) can be identified. These motifs are conserved amongst all covalent nucleotidyl transferases and are essential for catalysis. Although the crystal structure of DNA ligase IV has not been solved, structures have been solved for T7 ligase and for PBCV-1 capping enzyme.[49,50] Several excellent reviews detail these structures and they will not be discussed in detail here.[46,47,51] Together, these studies demonstrate that adenylation of the active site lysine acts as a conformational switch, facilitating rotation of the DNA–binding surface located in the OB domain towards the active site cleft, a feature that may enable only adenylated, and thereby active, ligases to form DNA complexes.[46] Interestingly, in distinction to other DNA ligases, DNA ligase IV exists endogenously as a pre-adenylated complex.[45] The significance of this is not clear but it may serve to facilitate the co-ordinated ligation of two DNA strands required for double stranded ligation.

Another feature of DNA ligase IV, in distinction to other DNA ligases, is the presence of two BRCT domains at the C-terminus . The C-terminus is required for interaction with XRCC4 although interaction requires the region between the two BRCT domains rather than the two domains themselves.[17,44] Interaction with XRCC4 enhances but is not essential for the adenylation activity of DNA ligase IV. However, XRCC4 is essential for DNA NHEJ and mice knocked out for either XRCC4 or DNA ligase IV share an identical phenotype with embryonic lethality occurring at the same developmental stage.[52-54] Cell lines defective in DNA ligase IV display genomic instability.[3] The crystal structure of human XRCC4 bound to a polypeptide corresponding to the interaction region of DNA ligase IV has been reported.[55] XRCC4 has a globular N-terminal head domain that folds into a seven-stranded, trumpet-shaped β-barrel with a long helical tai.[55,56] XRCC4 readily dimerizes via association of the two head domains. A tetrameric association of the protein has also been described, in which the two XRCC4 dimer tails interact. The sequence of DNA ligase IV that interacts with XRCC4 is a flat motif that folds into a portion of the helical tail. The residues at the interface between the two proteins are highly conserved across species suggesting that the interaction between the two proteins has been evolutionarily maintained. Unfortunately, the crystal structure provides little insight into the function of XRCC4 although it is clear that the interaction leaves the ligase domain of DNA ligase IV free to carry out its catalytic function in NHEJ.

In vitro assays have shown that Ku stimulates intermolecular DNA ligation by LX without affecting the joining of nicks.[32] The effect of Ku on ligation is observed with other mammalian DNA ligases and is not unique to DNA ligase IV. Since Ku is able to bridge two DNA molecules, the stimulation of ligation was attributed to Ku's ability to stabilize intermolecular association of DNA ends.[34] Subsequently, it has been shown that Ku also helps to recruit the LX complex to the DNA ends via a direct association between the proteins that is stimulated once Ku is bound to the DNA end.[57] Recruitment of LX by Ku causes inward translocation of Ku on the DNA molecule and requires a preadenylated LX molecule.[58] This "recruitment" role for Ku is specific for the LX complex, unlike its role in stimulating ligation.

Additional Proteins Implicated in DNA NHEJ

Artemis

To identify whether defects in NHEJ could contribute to human immunodeficiency, patients displaying severe combined immunodeficiency (SCID) were examined for their response to ionising radiation. One class of patient, designated radiosensitive SCID (RS-SCID), displayed radiosensitivity and impaired V(D)J recombination although activities of the known NHEJ proteins appeared to be intact.[59,60] The prevalence of RS-SCID amongst athabascan speaking Navajo Indians suggested a founder mutation in a NHEJ component which was finally identified via mapping studies as a new protein designated Artemis[61,62] (for a review see ref. 62). Artemis has, therefore, been designated as the sixth component of NHEJ. Cells lacking Artemis, however, display some differences to cells lacking the other NHEJ components. Most significantly, whilst cells lacking Ku, Xrcc4 and DNA ligase IV, are defective in the two rejoining steps involved in V(D)J recombination, namely signal and coding joint formation, cells lacking Artemis are impaired solely in coding joint formation.[61] Intriguingly, cells lacking DNA-PKcs are also defective only in coding joint formation.[64] The coding sequences are sub-exon fragments that constitute the immunoglobulin and T cell receptor loci, and the coding junctions thus contribute directly to the final V, D, or J segments. The signal sequences, on the other hand, serve to help target the site specific DSBs and the signal junctions do not contribute to the developing immune response. Intriguingly, as mentioned above, it had long been known that the coding ends form as hairpin intermediates that need to be resolved before coding join formation can take place.[65] Since the hairpins ends accumulate in cells lacking Ku or DNA-PKcs, it had long been argued that these proteins were required for their resolution although the mechanism remained elusive, with none of the component proteins or the DNA-PK complex possess hairpin cleavage activity. In a perceptive study, Ma et al,[66] showed that Artemis has a β-lactamase fold and can function as an exonuclease, whose activity becomes modified to include hairpin cleavage, upon phosphorylation by DNA-PKcs. This neatly explains the unique requirement of Artemis for coding join formation. In further contrast to the other NHEJ components, Artemis cells do not appear to display the characteristic defect in DNA DSB rejoining although they do display radiosensitivity.[59] The nuclease activities of Artemis may also be relevant in this context. Artemis has 5' to 3' processive exonuclease activity with single strand DNA specificity. This activity is also modified by phosphorylation such that the protein gains endonucleolytic cleavage activity rather than processive exonuclease activity. It has been suggested that Artemis may help to modify damaged DNA ends that might be refractory to direct rejoining.[63]

The Mre11/Rad50/Xrs2 (MRX) Complex

Studies in the genetically tractable yeast, *S. cerevisiae*, have implicated some additional proteins that function in, or regulate, NHEJ. The first of these is a complex of three proteins, Mre11, Rad50 and Xrs2, which has been designated the MRX complex.[67-70] *S. cerevisiae* deletion strains for any of these three proteins show impaired homologous recombination but strikingly, they are also defective in a plasmid based assay that monitors NHEJ, leading to the suggestion that the MRX complex functions during NHEJ, possibly to modify DNA ends since the complex has ssDNA endonuclease activity and ds 3'-5-exonuclease acitivities. This finding, however, does not appear to be reiterated in other organisms. In the fission yeast, *S. pombe*, strains lacking Rad32 (homologous to ScMre11) or Rad50 are proficient for NHEJ in an identical assay.[71] Furthermore, mammalian cells impaired for Nbs1, the functional homologue of the yeast Xrs2 protein, are proficient in their ability to undergo V(D)J recombination and do not display the characteristic DSB rejoining deficiency of NHEJ defective cells.[72,73] Notwithstanding this, they do display marked radiosensitivity similar to that displayed by NHEJ defective cells.[74] Similarly, cells deficient in Mre11 also appear to be proficient in NHEJ but display radiosensitivity.[75]

Sir2, 3 and 4 and Nej1p

The Sir2, 3 and 4 proteins have also been shown to be required for NHEJ in *S. cerevisiae*.[76] Subsequently, however, their impact of NHEJ was shown to be an indirect consequence of derepressing the *HMRa* and *HMLα* silent mating type genes.[77,78] Deletion of mating-type genes in a Sir^- haploid cell restored NHEJ to the same level seen in Sir^+ strains suggesting that mating type proteins controlled NHEJ. This was consolidated by the identification of a protein, designated Nej1p, that is transcriptionally repressed in Mata/Matα diploids.[79-82] Deletion of Nej1p reduced NHEJ 100-fold. Nej1p interacts with the N-terminus of Lif1 (the yeast homologue of XRCC4) and was reported to control its nuclear localization. Thus, in yeast, Nej1p appears to function to downregulate NHEJ in Mata/Matα diploid cells. Since diploid cells are able to carry out HR, this may serve as one of several mechanisms to regulate the two DSB rejoining processes, HR and NHEJ. To date a mammalian homologue of Nej1p has not been identified and whether NHEJ is similarly regulated is currently unknown.

Further Proteins Interacting with Core NHEJ Components

Some further interactions of NHEJ components with additional proteins that may function in the rejoining process have also been identified.

TdT

An important component that aids the antigen receptor diversity of the junctions during V(D)J recombination is terminal deoxynucleotidyl transferase (TdT). Consistent with a specific role in immune development, TdT expression is lymphoid specific.TdT has been shown to add nucleotides, designated N-nucleotides (i.e., non-templated-nucleotides), in a template-independent manner at the coding ends during V(D)J recombination.[83] Intriguingly, N-nucleotides were not found in the residual coding joints formed in Ku80 defective mice raising the possibility that Ku might be required for the recruitment of TdT.[84] More recently, a direct interaction has been demonstrated between Ku and TdT.[85] This interaction requires a conserved brct-like motif located at the N-terminus of TdT. Co-localization of Ku and TdT was also observed following the induction of DSBs in cells expressing TdT raising the possibility that Ku helps to localize TdT to the sites of V(D)J recombination.

Pol μ

Pol μ shows sequence similarity to TdT and, like TdT, has an amino-terminal BRCT domain.[86,87] Unlike TdT, however, Pol μ is expressed ubiquitously and is a template-dependent polymerase. This raised the possibility that Pol μ might function as a "fill-in" polymerase during NHEJ. This notion gained strength by the finding that agents that induce DSBs cause an increase in total cellular Pol μ levels and the localization of pol μ into discrete foci.[88] Additionally, Pol μ interacts with Ku, and requires the LX complex to form a stable interaction with Ku-bound DNA.[88] Finally, Pol μ aids the ability of LX and Ku to join ends with partially complementary overhangs. Whilst these in vitro and in vivo interaction studies are convincing, direct evidence for a role of Pol μ in NHEJ is still awaited.

Pol4

Another polymerase potentially linked to NHEJ is the *S. cerevisiae* polymerase, pol4. Pol4 is a member of the Pol X family of nucleotidyl transferases that includes TdT and Pol β, the polymerase that carries out gap filling during base excision repair.[89,90] Significantly, whilst Pol4 deficient yeast strains are proficient in rejoining linear plasmids with complementary single-strand ends, they are impaired in rejoining plasmids with noncomplementary termini.[91] Recently, additional evidence has emerged to consolidate a possible role for Pol4 in NHEJ. Pol4 can efficiently fill in short regions at a "gapped" double strand DNA end, an activity that was stimulated specifically by the yeast dnl4-Lif1 complex, the yeast homologue to the LX complex. Furthermore, Pol4 was shown to interact directly with Dnl4 via the Brct domain located at the N-terminus of Pol4.[92]

Werner Protein

Another protein that has been reported to interact with the Ku complex is the Werner protein.[93] Werner protein is defective in the human premature aging disorder, Werner's syndrome.[94] The WS gene, WRN, is a member of the REcQ family of helicases. WRNp has the anticipated helicase activity as well as associated DNA-dependent ATPase and exonuclease activities. One significant recent study demonstrated an interaction between WRNp and the Ku complex, and showed, moreover, that Ku strongly stimulated the 3'-5- exonuclease activity of WRNp.[94] Notwithstanding these interesting findings, WS cell lines do not display any significant radiosensitivity nor any marked defect in DSB repair. Additionally, no overt immunological defects have been reported in Werner's syndrome. Nonetheless, a role of WRN in a subset of NHEJ events cannot be ruled out. Alternatively, WRN may function together with Ku at telomeres.

Model for NHEJ and Potential Additional Interacting Proteins

The model (Fig. 1) is based on the data summarised in this article and represents a likely ordering of the events at the DNA end. The strong ds DNA end-binding properties of Ku suggests that it is likely to be a first step in the NHEJ process. In higher organisms, a second step of the reaction is likely to be the recruitment of DNA-PKcs. Interestingly, DNA-PKcs is

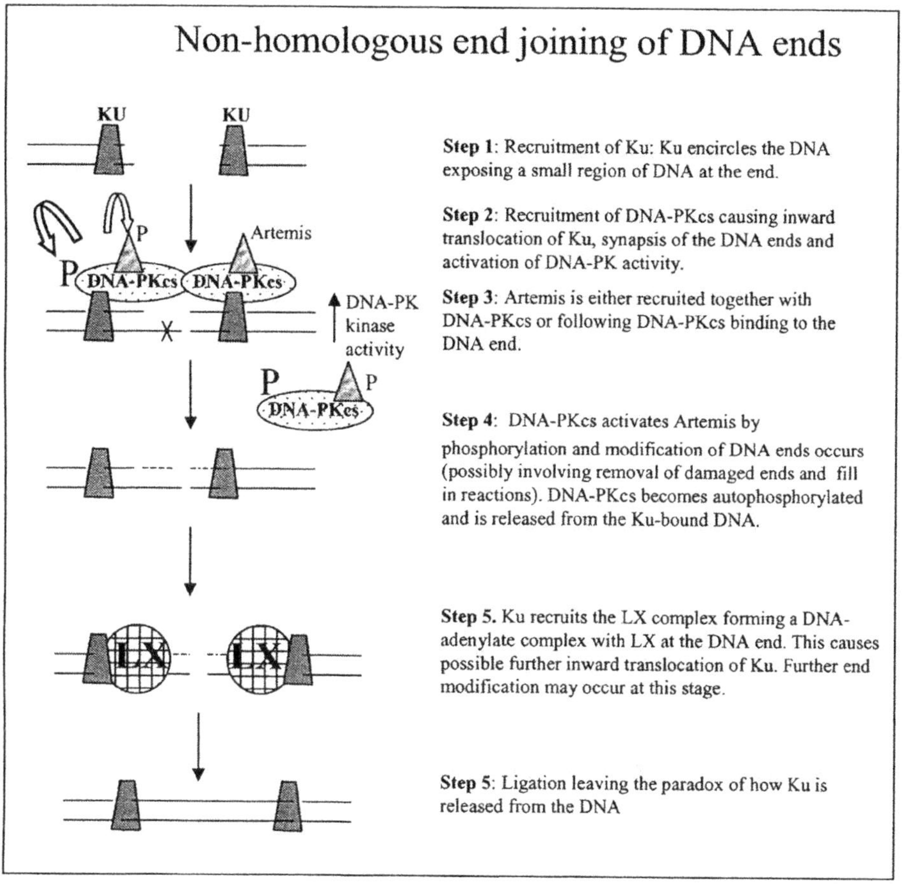

Figure 1. Non-homologous end joining of DNA.

dispensible for NHEJ in non-vertebrates which do not undergo V(D)J recombination. DNA-PKcs may, therefore, be a supplementary factor whose function is to recruit proteins, most notable Artemis, to faciliate the efficiency of V(D)J recombination. Nonetheless, DNA-PKcs is essential for NHEJ in higher organisms and has clearly evolved to be an essential component of the process. Intriguingly, some of the functions reported for DNA-PKcs can also be carried out by other NHEJ components. For example, DNA-PKcs has been postulated to enhance synapsis of the DNA, a function that has also been reported for Ku alone. Additionally, LX appears able to cause the same inward translocation of Ku that DNA-PKcs is able to achieve.[42,58] Prior to rejoining, it is likely that a range of end modification or processing steps will be necessary. The available evidence suggests that, at least one of these steps, the activation of Artemis, follows DNA-PK binding and activation. Thus, the DNA-PK complex may play an important role in recruiting and/or activating accessory proteins. Radiation induced breaks may also require end-modification to remove damaged bases. Fill-in by specific polymerases may also occur at this stage. The final step of NHEJ is the recruitment of LX by Ku. Since DNA-PKcs is a large complex, it is likely that dissociation of DNA-PKcs from the ends will be a pre-requisite for the recruitment of the LX complex and it is possibly significant that autophosphorylation of DNA-PKcs results in dissociation of the large protein from the DNA.[39]

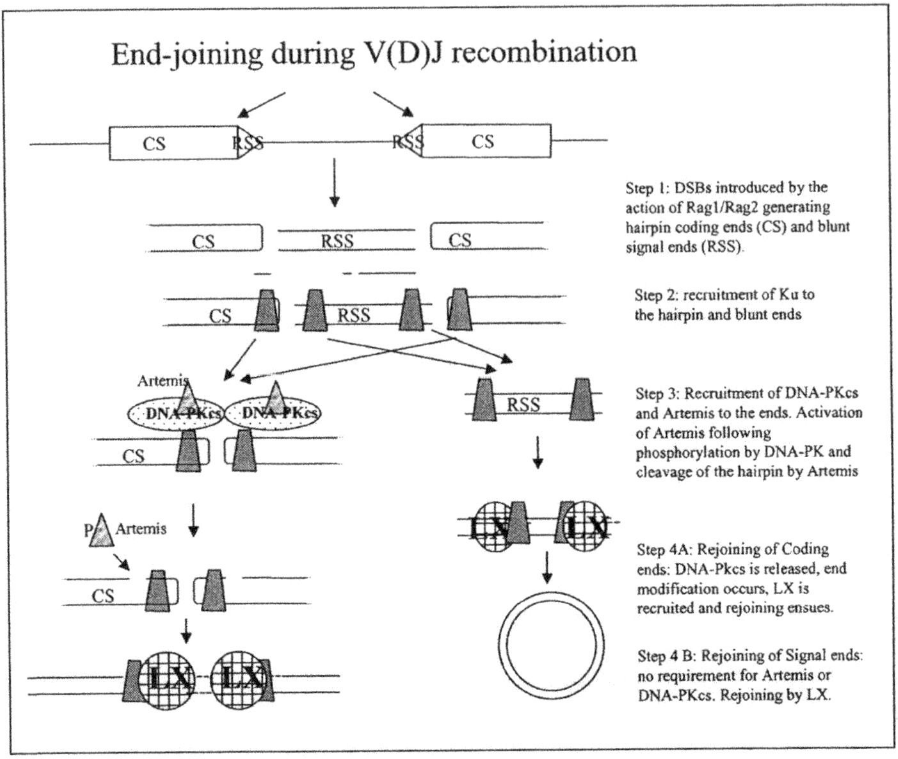

Figure 2. End-joining during V(D)J recombination.

This is the order shown in Figure 1 although it is not known whether or not LX is recruited to the DNA end with DNA-PKcs still bound. Double stranded ligation requires two co-ordinated ligation events. One potential function of XRCC4 is to enhance co-ordinated ligation possibly by forming a complex encompassing two ligase molecules. Alternatively or additionally, the LX complex may help to recruit further proteins involved in end-modification, a suggestion raised by the specific interaction between the yeast Pol4 and Dnl4 proteins.[92] The role of the NHEJ proteins in V(D)J recombination is shown in Figure 2. The notable feature in this case is that Artemis is required to resolve the hairpin intermediate generated at the coding junctions but is dispensable for the rejoining of signal junctions.

In conclusion, a range of biochemical, biophysical and cellular studies have contributed to provide significant insight into the process of NHEJ so that detailed models can now be proposed. However, there is still much to be learned and most strikingly, there appears to be at least some further proteins to be identified that either function in or regulate the process.[95]

Acknowledgements

I would like to thank Dr. M. O'Driscoll and E. Riballo for helpful discussions and careful reading of the manuscript. The PAJ laboratory is supported by grants from the Medical Research Council, the Leukaemia Research Fund, the Primary Immunodeficiency Association, the Human Frontiers Science Programme, the Department of Health and the European Union.

References

1. Jeggo PA. DNA breakage and repair. Adv Genet 1998; 38:185-211.
2. Kanaar R, Hoeijmakers JH, van Gent DC. Molecular mechanisms of DNA double strand break repair. Trends Cell Biol 1998; 8:483-489.
3. Karanjawala ZE, Grawunder U, Hsieh CL et al. The nonhomologous DNA end joining pathway is important for chromosome stability in primary fibroblasts. Curr Biol 1999; 9:1501-1504.
4. Taccioli GE, Rathbun G, Oltz E et al. Impairment of V(D)J recombination in double-strand break repair mutants. Science 1993; 260:207-210.
5. Gellert M. V(D)J recombination: RAG proteins, repair factors, and regulation. Annu Rev Biochem 2002; 71:101-132.
6. O'Driscoll M, Jeggo P. Immunological disorders and DNA Repair. Mutation Res 2002; 1-18.
7. Rolink A, Melchers F, Andersson J. The SCID but not the RAG-2 gene product is required for S mu-S epsilon heavy chain class switching. Immunity 1996; 5:319-330.
8. Casellas R, Nussenzweig A, Wuerffel R et al. Ku80 is required for immunoglobulin isotype switching. EMBO J 1998; 17:2404-2411.
9. Manis JP, Gu Y, Lansford R et al. Ku70 is required for late B cell development and immunoglobulin heavy chain class switching. J Exp Med 1998; 187:2081-2089.
10. Boulton SJ, Jackson SP. Components of the Ku-dependent non-homologous end-joining pathway are involved in telomeric length maintenance and telomeric silencing. EMBO J 1998; 17:1819-1828.
11. Taccioli GE, Gottlieb TM, Blunt T et al. Ku80: product of the XRCC5 gene. Role in DNA repair and V(D)J recombination. Science 1994; 265:1442-1445.
12. Smider V, Rathmell WK, Lieber MR et al. Restoration of X-ray resistance and V(D)J recombination in mutant-cells by Ku cDNA. Science 1994; 266:288-291.
13. Dvir A, Peterson SR, Knuth MW et al. Ku autoantigen is the regulatory component of a template-associated protein kinase that phosphorylates RNA polymerase II. Proc Natl Acad Sci USA 1992; 89:11920-11924.
14. Gottlieb TM, Jackson SP. The DNA-dependent protein kinase: requirement of DNA ends and association with Ku Antigen. Cell 1993; 72:131-142.
15. Blunt T, Finnie NJ, Taccioli GE et al. Defective DNA-dependent protein kinase activity is linked to V(D)J recombination and DNA repair defects associated with the murine scid mutation. Cell 1995; 80:813-823.
16. Li Z, Otevrel T, Gao Y et al. The XRCC4 gene encodes a novel protein involved in DNA double-strand break repair and V(D)J recombination. Cell 1995; 83:1079-1089.
17. Critchlow SE, Bowater RP, Jackson SP. Mammalian DNA double-strand break repair protein XRCC4 interacts with DNA ligase IV. Curr Biol 1997; 7:588-598.
18. Grawunder U, Zimmer D, Fugmann S et al. DNA ligase IV is essential for V(D)J recombination and DNA double-strand break repair in human precursor lymphocytes. Mol Cell 1998; 2:477-484.
19. Whitehouse CJ, Taylor RM, Thistlethwaite A Z et al. XRCC1 stimulates human polynucleotide kinase activity at damaged DNA termini and accelerates DNA single-strand break repair. Cell 2001; 104:107-117.
20. Chappell C, Hanakahi LA, Karimi-Busheri F et al. Involvement of human polynucleotide kinase in double-strand break repair by non-homologous end joining. EMBO J 2002; 21:2827-2832.
21. Oettinger MA, Schatz DG, Gorka C et al. RAG-1 and RAG-2, adjacent genes that synergistically activate V(D)J recombination. Science 1990; 248:1517-1523.
22. McBlane JF, van Gent DC, Ramsden DA et al. Cleavage at a V(D)J recombination signal requires only RAG1 and RAG2 proteins and occurs in two steps. Cell 1995; 83:387-395.
23. Blier PR, Griffith AJ, Craft J et al. Binding of Ku protein to DNA. Measurement of affinity for ends and demonstration of binding to nicks. J Biol Chem 1993; 268:7594-7601.
24. de Vries E, van Driel W, Bergsma WG et al. HeLa nuclear protein recognizing DNA termini and translocating on DNA forming a regular DNA-multimeric protein complex. J Mol Biol 1989; 208:65-78.
25. Smith GC, Jackson SP. The DNA-dependent protein kinase. Genes Dev. 1999; 13:916-934.
26. Walker JR, Corpina RA, Goldberg J. Structure of the Ku heterodimer bound to DNA and its implications for double-strand break repair. Nature 2001; 412:607-614.

27. Dynan WS, Hoo S. Interaction of Ku protein and DNA-dependent protein kinase catalytic subunit with nucleic acids. Nucl Acids Res 1998; 26:1551-1559.
28. Singleton BK, Torres-Arzayus MI, Rottinghaus ST et al. The C terminus of Ku80 activates the DNA-dependent protein kinase catalytic subunit. Mol Cell Biol 1999; 19:3267-3277.
29. Gell D, Jackson SP. Mapping of protein-protein interactions within the DNA-dependent protein kinase complex. Nucleic Acids Res 1999; 27:3494-3502.
30. Singleton BK, Priestley A, Gell D et al. Molecular and biochemical characterisation of mutants defective in Ku80. Mol Cell Biol 1997; 17:1264-1273.
31. Arosio D, Cui S, Ortega C et al. Studies on the mode of Ku interaction with DNA. J Biol Chem 2002; 16:9741-9748.
32. Ramsden DA, Gellert M. Ku protein stimulates DNA end joining by mammalian DNA ligases: a direct role for Ku in repair of DNA double-strand breaks. EMBO J 1998; 17:609-614.
33. Cary RB, Peterson SR, Wang J et al. DNA looping by Ku and the DNA-dependent protein kinase. Proc Natl Acad Sci USA 1997; 94:4267-4272.
34. Bliss TM, Lane DP. Ku selectively transfers between DNA molecules with homologous ends. J Biol Chem 1997; 272:5765-5773.
35. Blunt T, Gell D, Fox M et al. Identification of a nonsense mutation in the carboxy-terminal region of DNA-dependent protein kinase catalytic subunit in the scid mouse. Proc Natl Acad Sci USA 1996; 93:10285-10290.
36. Beamish HJ, Jessberger R, Riballo E et al. The C-terminal conserved domain of DNA-PKcs, missing in the SCID mouse, is required for kinase activity. Nucleic Acids Res 2000; 28:1506-1513.
37. Kurimasa A, Kumano S, Boubnov N et al. Requirement for the kinase activity of human DNA-dependent protein kinase catalytic subunit in DNA strand break rejoining. Mol Cell Biol 1999; 19:3877-3884.
38. Chan DW, Chen BP, Prithivirajsingh S et al. Autophosphorylation of the DNA-dependent protein kinase catalytic subunit is required for rejoining of DNA double-strand breaks. Genes Dev 2002; 16:2333-2338.
39. Merkle D, Douglas P, Moorhead GB et al. The DNA-dependent protein kinase interacts with DNA to form a protein-DNA complex that is disrupted by phosphorylation. Biochem 2002; 41:12706-12714.
40. Yaneva M, Kowalewski T, Leiber MR. Interaction of DNA-dependent protein kinase with DNA and with Ku: biochemical and atomic-force microscopy studies. EMBO J 1997; 16:5098-5112.
41. Hammarsten O, Chu G. DNA-dependent protein kinase: DNA binding and activation in the absence of Ku. Proc Natl Acad Sci USA 1998; 95:525-530.
42. Yoo S, Dynan WS. Geometry of a complex formed by double strand break repair proteins at a single DNA end: recruitment of DNA-PKcs induces inward translocation of Ku protein. Nucleic Acid Res 1999; 27:4679-4686.
43. DeFazio LG, Stansel RM, Griffith JD et al. Synapsis of DNA ends by DNA-dependent protein kinase. EMBO J 2002; 21:3192-3200.
44. Grawunder U, Wilm M, Wu X et al. Activity of DNA ligase IV stimulated by complex formation with XRCC4 protein in mammalian cells. Nature 1997; 388:492-495.
45. Wei Y-F, Robins P, Carter K et al. Molecular cloning and expression of human cDNAs encoding a novel DNA ligase IV and DNA ligase III, an enzyme active in DNA repair and recombination. Mol Cell Biol 1995; 15:3206-3216.
46. Doherty AJ, Suh SW. Structural and mechanistic conservation in DNA ligases. Nucleic Acids Res 2000; 28:4051-4058.
47. Martin IV, MacNeill SA. ATP-dependent DNA ligases. Genome Biol 2002; 3:REVIEWS3005.
48. Bycroft M, Hubbard TJ, Proctor M et al. The solution structure of the S1 RNA binding domain: a member of an ancient nucleic acid-binding fold. Cell 1997; 88:235-242.
49. Odell M, Sriskanda V, Shuman S et al. Crystal structure of eukaryotic DNA ligase-adenylate illuminates the mechanism of nick sensing and strand joining. Mol Cell 2000; 6:1183-1193.
50. Subramanya HS, Doherty AJ, Ashford SR et al. Crystal structure of an ATP-dependent DNA ligase from bacteriophage T7. Cell 1996; 85:607-615.
51. Timson DJ, Singleton MR, Wigley DB. DNA ligases in the repair and replication of DNA. Mutat Res 2000; 460:301-318.

52. Frank KM, Sekiguchi JM, Seidl KJ et al. Late embryonic lethality and impaired V(D)J recombination in mice lacking DNA ligase IV. Nature 1998; 396:173-7.
53. Gao Y, Sun Y, Frank KM et al. A critical role for DNA end-joining proteins in both lymphogenesis and neurogenesis. Cell 1998; 95:891-902.
54. Barnes DE, Stamp G, Rosewell I et al. Targeted disruption of the gene encoding DNA ligase IV leads to lethality in embryonic mice. Curr Biol 1998; 8:1395-1398.
55. Sibanda BL, Critchlow SE, Begun J et al. Crystal structure of an Xrcc4-DNA ligase IV complex. Nat Struct Biol 2001; 8:1015-1019.
56. Junop MS, Modesti M, Guarne A et al. Crystal structure of the Xrcc4 DNA repair protein and implications for end joining. EMBO J 2000; 19:5962-5970.
57. Nick McElhinny SA, Snowden CM, McCarville J et al. Ku recruits the XRCC4-ligase IV complex to DNA ends. Mol Cell Biol 2000; 20:2996-3003.
58. Kysela B, Doherty AJ, Chovanec M et al. Ku stimulation of DNA ligase IV-dependent ligation requires inward movement along the DNA molecule. J Biol Chem 2003; In press:
59. Nicolas N, Finnie NJ, Cavazzana-Calvo M et al. Lack of detectable defect in DNA double-strand break repair and DNA- dependent protein kinase activity in radiosensitive human severe combined immunodeficiency fibroblasts. Eur J Immunol 1996; 26:1118-1122.
60. Nicolas N, Moshous D, Cavazzana-Calvo M et al. A human severe combined immunodeficiency (SCID) condition with increased sensitivity to ionizing radiations and impaired V(D)J rearrangements defines a new DNA recombination/repair deficiency. J Exp Med 1998; 188:627-634.
61. Moshous D, Callebaut I, de Chasseval R et al. Artemis, a novel DNA double-strand break repair/ V(D)J recombination protein, is mutated in human severe combined immune deficiency. Cell 2001; 105:177-186.
62. Moshous D, Li L, Chasseval R et al. A new gene involved in DNA double-strand break repair and V(D)J recombination is located on human chromosome 10p. Hum Mol Genet 2000; 9:583-588.
63. Jeggo PA, O'Neill P. The Greek goddess, Artemis, reveals the secrets of her cleavage. DNA Repair 2002; 67:1-7.
64. Lieber MR, Hesse JE, Lewis S et al. The defect in murine severe combined immune deficiency: joining of signal sequences but not coding segments in V(D)J recombination. Cell 1988; 55:7-16.
65. Gellert M. Recent advances in understanding V(D)J recombination. Adv Immunol 1997; 64:39-64.
66. Ma Y, Pannicke U, Schwarz K et al. Hairpin Opening and Overhang Processing by an Artemis/ DNA-Dependent Protein Kinase Complex in Nonhomologous End Joining and V(D)J Recombination. Cell 2002; 108:781-794.
67. Milne GT, Jin S, Shannon KB et al. Mutations in two Ku homologs define a DNA end-joining repair pathway in Saccharomyces cerevisiae. Mol Cell Biol 1996; 16:4189-4198.
68. Moore JK, Haber JE. Cell cycle and genetic requirements of two pathways of nonhomologous end-joining repair of double-strand breaks in Saccharomyces cerevisiae. Mol Cell Biol 1996; 16:2164-2173.
69. Tsukamoto Y, Kato J, Ikeda H. Effects of mutations of RAD50, RAD51, RAD52, and related genes on illegitimate recombination in Saccharomyces cerevisiae. Genetics 1996; 142:383-391.
70. Schiestl RH, Zhu J, Petes TD. Effect of mutations in genes affecting homologous recombination on restriction enzyme-mediated and illegitimate recombination in Saccharomyces cerevisiae. Mol Cell Biol 1994; 14:4493-4500.
71. Manolis KG, Nimmo ER, Hartsuiker E et al. Novel functional requirements for non-homologous DNA end joining in Schizosaccharomyces pombe. EMBO J 2001; 20:1-12.
72. Girard P-M, Foray N, Stumm M et al. Radiosensitivity in Nijmegen Breakage Syndrome cells is due to a repair defect and not cell cycle checkpoint defects. Cancer Res 2000; 60:4881-4888.
73. Yeo TC, Dong X, Sabath DE et al. V(D)J Rearrangement in Nijmegen Breakage Syndrome. Mol Immunol 2000; 37:1131-1139.
74. Shiloh Y. Ataxia-telangiectasia and the Nijmegen breakage syndrome: related disorders but genes apart. Annu Rev Genet 1997; 31:635-662.
75. Stewart GS, Maser RS, Stankovic T et al. The DNA double-strand break repair gene hMRE11 is mutated in individuals with an ataxia-telangiectasia-like disorder. Cell 1999; 99:577-587.
76. Tsukamoto Y, Kato J, Ikeda H. Silencing factors participate in DNA repair and recombination in Saccharomyces cerevisiae. Nature 1997; 388:900-903.

77. Astrom SU, Okamura SM, Rine J. Yeast cell-type regulation of DNA repair. Nature 1999; 397:310.
78. Lee SE, Paques F, Sylvan J et al. Role of yeast SIR genes and mating type in directing DNA double-strand breaks to homologous and non-homologous repair paths. Curr Biol 1999; 9:767-770.
79. Kegel A, Sjostrand JO, Astrom SU. Nej1p, a cell type-specific regulator of nonhomologous end joining in yeast. Curr Biol 2001; 11:1611-1617.
80. Frank-Vaillant M, Marcand S. NHEJ regulation by mating type is exercised through a novel protein, Lif2p, essential to the Ligase IV pathway. Genes Dev 2001; 15:3005-3012.
81. Valencia M, Bentele M, Vaze MB et al. NEJ1 controls non-homologous end joining in Saccharomyces cerevisiae. Nature 2001; 414:666-669.
82. Ooi SL, Shoemaker DD, Boeke JD. A DNA Microarray-Based Genetic Screen for Nonhomologous End-Joining Mutants in Saccharomyces cerevisiae. Science 2001; 8:
83. Gauss GH, Lieber MR. Mechanistic constraints on diversity in human V(D)J recombination. Mol Cell Biol 1996; 16:258-269.
84. Bogue MA, Wang C, Zhu C et al. V(D)J recombination in Ku86-deficient mice: distinct effects on coding, signal, and hybrid joint formation. Immunity 1997; 7:37-47.
85. Mahajan KN, Gangi-Peterson L, Sorscher DH et al. Association of terminal deoxynucleotidyl transferase with Ku. Proc Natl Acad Sci USA 1999; 96:13926-13931.
86. Aoufouchi S, Flatter E, Dahan A et al. Two novel human and mouse DNA polymerases of the polX family. Nucleic Acids Res 2000; 28:3684-3693.
87. Dominguez O, Ruiz JF, Lain De Lera T et al. DNA polymerase mu (Pol **m**) , homologous to TdT, could act as a DNA mutator in eukaryotic cells. EMBO J 2000; 19:1731-1742.
88. Mahajan KN, Nick McElhinny SA, Mitchell BS et al. Association of DNA polymerase mu (pol mu) with Ku and ligase IV: role for pol mu in end-joining double-strand break repair. Mol Cell Biol 2002; 22:5194-5202.
89. Shimizu K, Santocanale C, Ropp PA et al. Purification and characterization of a new DNA polymerase from budding yeast Saccharomyces cerevisiae. A probable homolog of mammalian DNA polymerase beta. J Biol Chem 1993; 268:27148-27153.
90. Prasad R, Widen SG, Singhal RK et al. Yeast open reading frame YCR14C encodes a DNA beta-polymerase-like enzyme. Nucleic Acids Res 1993; 21:5301-5307.
91. Wilson TE, Lieber MR. Efficient processing of DNA ends during yeast nonhomologous end joining. Evidence for a DNA polymerase beta (Pol4)-dependent pathway. J Biol Chem 1999; 274:23599-23609.
92. Tseng HM, Tomkinson AE. A physical and functional interaction between yeast Pol4 and Dnl4-Lif1 links DNA synthesis and ligation in nonhomologous end joining. J Biol Chem 2002; 277:45630-45637.
93. Cooper MP, Machwe A, Orren DK et al. Ku complex interacts with and stimulates the Werner protein. Genes Dev 2000; 14:907-912.
94. Yu C-E, Oshima J, Fu Y-H et al. Positional cloning of the Werner's syndrome gene. Science 1996; 272:258-262.
95. Dai Y, Kysela B, Hanakahi LA et al. Non-homologous end joining and V(D)J recombination require an additional factor. Proc Natl Acad Sci USA 2003; 100:2462-2467.

Index

A

Alkylating agents 15, 30, 33-36, 50, 52, 55, 57, 58, 66, 90, 101, 102, 125
AP endonuclease 27, 30-33, 37, 39, 52-54
AP endonuclease-1 (APE1) 29-31, 35-38, 51-54
Apoptosis-inducing factor (AIF) 50
Artemis 150, 151, 153, 154
Ataxia telangiectasia (AT) 75, 110, 112, 116-118, 120, 122
Ataxia telangiectasia mutated (ATM) 3, 4, 7, 12, 14, 19, 20, 25, 26, 75, 76, 78, 101-103, 107, 110-112, 114, 116-120, 122, 123, 125-128, 145
Ataxia telangiectasia related (ATR) 3, 9, 10, 12, 14, 16-18, 75, 76, 78, 89, 107, 110, 117, 122, 123, 125-128, 130
ATR-interacting protein (ATRIP) 3, 9, 10, 12, 17, 123, 127, 128

B

53BP1 2, 107, 110-112, 114, 116-118, 120, 123, 126, 129, 130
Base excision repair (BER) 17, 27-39, 49-57, 64, 66, 100, 122, 131, 152
BLM 3, 101, 107, 111, 123, 125, 129, 130
Bloom's Syndrome (BS) 101, 129, 130
BRCA1 2, 6, 17, 20, 101, 107, 110, 111, 114, 116-121, 123, 126, 129
BRCA2 107, 111, 119-121
BRCT (*BRCA1* carboxyl terminus) 2, 3, 6, 34, 35, 52, 55, 107, 114, 117, 120, 129, 149, 152
Budding yeast 1, 3, 6-10, 12, 14-19, 114

C

Caffeine 65, 76, 117
Carboplatin 90
Cell cycle 1-7, 10, 14-18, 20, 30, 35, 47, 57, 64, 66, 74, 75, 91, 101, 103, 107, 108, 110, 111, 114, 120, 126, 128
Cell cycle arrest 1, 5, 6, 15, 16, 74, 91
Chk1 3, 5, 6, 10, 14, 110, 111, 114, 117, 122, 123, 126-128
Chk2 3, 75, 110, 111, 114, 116, 117, 122, 127, 128
Chromatin 7, 9, 11-13, 16-19, 50, 68, 72, 74, 78, 110-112, 114, 115, 118, 126-130
Chromosome segregation 1
Cisplatin 90, 102, 103, 125
Clamp and clamp-loader complex 8, 107
Cockayne syndrome (CS) 63-65, 68, 70, 72, 73, 78, 79
Cyclin-dependent kinase (CDK) 10

D

DDB2 63, 65, 66, 68, 73, 74, 76
DDC1 7
Ddc2 3, 5, 6, 9, 10-13, 17, 127
Defl 73
DNA
- damage 1-8, 10-20, 27, 28, 35, 36, 49, 52, 56, 63, 65, 66, 68-72, 75, 76, 78, 90, 101-103, 110, 121, 125-127
- damage checkpoint 1, 4, 6, 7, 10, 14-19
- glycosylase 27-30, 36, 39, 51, 53
- ligase 14, 27, 33-35, 37-39, 53-56, 72, 91, 146, 147, 149, 150
- ligase III 33-35, 37, 39, 53, 55
- ligase IV 146, 147, 149, 150
- polymerase 2, 13, 15, 27, 31-33, 38, 50, 53-55, 65, 76, 90, 100
- polymerase β (Polβ) 27, 31, 32, 51-56
- replication 1, 4, 6, 8, 13, 15, 17, 30, 32, 33, 49, 50, 57, 58, 63-66, 72, 73, 75, 76, 78, 90, 92, 101-103, 107-109, 113, 119-121, 124, 126, 129, 130
- replication fork 15, 75, 76, 108, 109, 113, 129

DNA-PK 110, 113, 118, 119, 123, 125, 147-150, 153
DNA-PKcs 110, 116, 118-120, 146-150, 152-154
Double strand break repair (DSBR) 64, 66
Double-strand break (DSB) 2-4, 7, 9, 12-20, 49, 50, 52, 55-58, 107-130, 146, 147, 149-152
Drug resistance 101-103

E

E2F1 68
Excision repair 4, 17, 27, 28, 31-33, 36, 49, 54, 63, 64, 66, 68, 74, 100, 108, 122, 152
Exonuclease 1 (EXO1) 98

F

Flap endonuclease 1 (FEN1) 35, 38, 39, 51, 53, 54, 58
Fluorescence recovery after photobleaching (FRAP) 18
'Futile Repair Cycle' model 101

G

γH2AX 107, 108, 110-118, 121, 125, 126, 128-130, 132
GADD45 66, 74
Gene transcription 1, 65, 66
Genome surveillance 1, 16, 101
G-proteins 92

H

H2AX 12, 18-20, 107, 110-117, 123, 128, 129
HAP1 30
HDAC4 111, 116
Helicase 3, 15, 65, 70, 71, 91, 101, 107, 122, 123, 129, 130, 152
Hereditary Non-Polyposis Colorectal Cancer (HNPCC) 90, 93, 98, 99
HIPK2 75
HRR 108, 111, 117, 119-122, 125, 128

I

Intra-S checkpoint 2, 3, 14-16

J

JNK 102, 103

K

Kinase 3-6, 10, 12-14, 16, 17, 19, 20, 34, 35, 54, 69, 70, 75, 76, 78, 103, 107, 110, 112, 116, 118, 122, 123, 125-127, 146-149
Ku 13, 17, 146-150, 152, 153

L

Long-patch pathway 27, 33, 34, 38, 39

M

MBD4 28, 29, 100
Mecl 3-7, 9-13, 15-19, 125, 127
MEC3 7
MED1 100
Mediator of DNA checkpoint (MDC1) 107, 110-114, 116, 118, 122, 130
see also NFBD1
Methyl-CpG binding endonuclease I 100
Microsatellite instability (MSI) phenotype 90
Mismatch repair (MMR) 32, 33, 64, 65, 68, 69, 90-93, 96-103
MLH1 69, 90-93, 95, 96, 98-103
MLH3 91-93, 95, 96
'MMR Dependent Replication Stalling' model 102
Mre11 complex 3, 4, 12-15, 17, 75, 78, 128
Mre11-Rad50-Nbs1 complex (MRN) complex 20, 107, 112, 113, 115, 117-122, 125, 130
Mre11/Rad50/Xrs2 (MRX) complex 14, 151
MSH2 69, 90-92, 96, 98-103
MSH3 91, 92, 96, 98-100
MSH6 91, 92, 96, 98, 100, 101
MutL 90-96, 101, 103
MutS 90-93, 96-98, 100, 101, 103

N

Nej1p 151
NFBD1 110, 111, 114, 116 *see also* MDC1
Nijmegen breakage syndrome (NBS) 14, 75, 113
Non-homologous end-joining (NHEJ) 13-15, 19, 108, 110, 111, 118, 120, 122, 125, 128, 146-154
Nuclear foci 17, 107, 110, 112-116, 126, 128
Nucleotide excision repair (NER) 4, 16-18, 32, 33, 63-76, 78, 79

O

OGG1 28-30, 37, 38

P

P127 65, 68, 73
P53 19, 63, 66-68, 70, 74, 75, 78, 102, 103, 126
Phosphorylation 5-7, 9, 10, 12, 14, 16, 18, 19, 30, 66, 69, 70, 72, 73, 75, 76, 98, 103, 107, 109-114, 116-119, 121, 123, 125-128, 130, 149-151, 153
PI3K-like kinase (PIKK) 3, 6, 7, 9, 10, 12, 127, 148, 149
PMS1 91-93, 95, 96
PMS2 69, 91-93, 95, 96, 99, 101, 103
Pol4 152, 154
Pol D 72
Pol E 72
Pol μ 152
Poly(ADP-ribose)polymerase (PARP-1) 34-36, 50-53, 55, 56
Polymerase 2, 8, 12, 13, 15, 27, 30-33, 35, 36, 38, 50, 52-55, 63, 65, 72, 73, 76-78, 90, 92, 100, 121, 122, 128, 152, 153
Polynucleotide kinase (PNK) 34, 35, 51, 54, 56, 147
Proliferating cell nuclear antigen (PCNA) 2, 5, 7-9, 11-13, 17, 30, 33-35, 38, 39, 51, 54, 55, 58, 65, 72, 74, 75, 77, 78, 92, 100, 101, 103, 123, 127-130
Purine analogues 90

R

Rad3 2-4, 10, 12, 110, 125, 127, 151
RAD9 6
RAD17 7
RAD24 6, 7, 12
Rad51 14, 17, 18, 107, 111-121, 125
Rad53 2, 3, 5-7, 10, 12, 14, 15, 17
Radioresistant DNA synthesis 122
Reactive oxygen species (ROS) 27, 31, 49, 50, 147
Recombination bypass 102
Refl 30
Replication factor C (RFC) 2, 5, 7-9, 12, 15, 72, 77, 101, 127
Replication fork collapse 1
Replication protein A (RPA) 6, 12, 15, 30, 34, 65, 67, 68, 70-72, 76, 111, 119, 120, 123, 126
Retinoids 72
Ribonucleotide reductase (RNR) 10, 15
RNA pol II 63, 65, 68-70, 72-75
Rsp5 73

S

Severe combined immunodeficiency (SCID) 146-150
Sgs1 3, 15
Short-patch pathway 27, 39
Single-strand breaks (SSB) 12, 31, 35, 49-58, 107-109, 120-122, 147
Sir2, 3 and 4 151
Small ubiquitin-related modifying enzyme (SUMO) 78
SMUG1 28-30
Ssa1 and Ssa2 7

T

6- thioguanine 90
TdT 151, 152
Tel1 2, 3, 6, 10
TFIIH 65, 69-73
TFIIS 72
The Mre11/Rad50/Xrs2 (MRX) complex 151
Thyroid hormone 72
TopBP1 3, 111, 120, 123, 129
Transcription 1, 5, 10, 18, 30, 31, 49, 50, 63-70, 72-75, 78, 109, 117, 121, 125, 151
Transcription coupled repair (TCR) 63, 65, 66, 68-70, 72, 73
Transmission electron microscopy (TEM) 8
Trichothiodystrophy (TTD) 64

U

Ubiquitin ligase 73
Ubiquitination 63, 65, 67, 69, 73-76, 78
UNG family 28
UNG2 29, 30
UV irradiation 4, 7, 9, 31, 49, 63, 64, 67, 68, 70, 73-76, 112, 122, 125
UV photoproducts 16, 66, 70

W

Werner protein 35, 152
Werner's syndrome 107, 129, 152

X

Xeroderma pigmentosum (XP) 4, 64, 65, 68, 70, 72, 76, 78, 79, 122, 125, 128, 129
Xeroderma pigmentosum group A (XPA) 4, 65, 67, 70-72, 78
XRCC1 33-35, 37, 39, 49-58, 147
XRCC4 110, 146, 147, 149-151, 154

MIX
Papier aus verantwortungsvollen Quellen
Paper from responsible sources
FSC® C105338

Printed by Books on Demand, Germany